Aeroponics

Aeroponics
Growing Vertical

Thomas W. Gurley

CRC Press
Taylor & Francis Group
Boca Raton London New York

CRC Press is an imprint of the
Taylor & Francis Group, an **informa** business

First edition published 2020
by CRC Press
6000 Broken Sound Parkway NW, Suite 300, Boca Raton, FL 33487-2742

and by CRC Press
2 Park Square, Milton Park, Abingdon, Oxon, OX14 4RN

© 2020 Taylor & Francis Group, LLC

CRC Press is an imprint of Taylor & Francis Group, LLC

ISBN: 978-0-367-40953-1 (hbk)
ISBN: 978-0-367-37430-3 (pbk)
ISBN: 978-0-367-81007-8 (ebk)

Typeset in Times
by codeMantra

Contents

Preface

Vertical column aeroponics is a relatively new variation of hydroponics which initially was practiced by misting the roots of a suspended plant with nutrient solution. However, hydroponics has been practiced for several decades and is defined as a method of growing plants with no soil (in some cases solid media) with their roots completely submerged in the nutrient solution. Both of these techniques, aeroponics and hydroponics, are considered to be soil-less agriculture and more recently have been designated as controlled environment agriculture (CEA). Both of these approaches can be practiced as what some people referred to as "vertical farming." The difference is that the new version of aeroponics is conducted with vertical columns with plants growing one above another, whereas "vertical" hydroponics can be conducted by stacking horizontal trays of the nutrient solution one above the next so they are stacked vertically. The two major differences between vertical column aeroponics and vertical farming with hydroponics is that: first, with aeroponics sunlight can be utilized in a greenhouse environment, whereas it is necessary to utilize artificial light in the hydroponic case due to the stacking requirement. Second, with aeroponics the roots are in direct contact with air and specifically oxygen, whereas with hydroponics the roots only come in contact with the dissolved oxygen in water. Oxygen is extremely insoluble in water, so these plants are exposed to much lower concentrations of oxygen as compared to the roots in direct contact with air.

This book explores the history of the development of aeroponics which is rooted in the fundamental discoveries associated with botany and chemistry and is directly related to the development of hydroponics.

The book attempts to clearly define what aeroponics really is and to point out the similarities and differences with concomitant hydroponic growing methods. Chapter 3 deals with the value proposition for aeroponics.

The science of aeroponics is presented in Chapter 4 which reviews all the technical peer-reviewed papers in the scientific literature since the 1970s. Over 200 papers are summarized and the scope of research, the variety of research, and the geographical diversity of the research are all highlighted. The number of papers presented each year since 1970 has increased exponentially which seems to show that this technique has some traction and will continue to make a valuable contribution to CEA in the future. Research has been conducted in all the major geographical areas of the world as well as in space.

Innovation of aeroponics is presented in Chapter 5 which looks at a similar trend in the patent literature which compliments what has been observed in the scientific literature. This demonstrates not only the scientific/engineering value of aeroponics but also the economic value.

Chapter 6 covers the business of aeroponics and describes the entrepreneurial activity in this area from both manufacturers of aeroponic growing systems as well as some of the key growers using this approach. One of the latest developments is the construction of "indoor farms" located in shipping containers and some of those companies are also highlighted.

The practice of aeroponics is presented in Chapter 7 and that covers the key aspects of this type of technology requirements—nutrients, water quality, and many practical aspects.

Chapter 8 provides a few current examples of research using vertical column aeroponics is presented from work conducted at Rutgers University using commercial systems and at Charleston Southern University and the Citadel using R&D vertical growing systems. The focus of these research efforts is to better understand how to optimize these systems for the best yields and the concomitant nutrient requirements.

This book is written from the perspective of a practitioner, a scientist, and a business person for the sole purpose of compiling information from the accumulated knowledge that has been published on this newly emerging technology—aeroponics. It is my desire that this book would impact people in all spheres of society—science, practice, business, education, and the general public—to open everyone's eyes to see the potential of this methodology for the production of pure, clean, and safe food for the future. It can also be used as a textbook for an introductory course in aeroponics. In addition, this book also demonstrates the many benefits of growing food in this way that is efficient, effective, and sustainable. This is a book of answers for one of the most challenging questions that face our world today—how are we going to feed everyone in the 21st millennium and beyond?

Author

Thomas W. Gurley is an adjunct professor of chemistry at Charleston Southern University. He was a Fulbright Scholar and Fulbright Specialist in Ukraine at the National Academy of Science, Institute of Single Crystals and also in Uganda at Uganda Christian University—Agricultural Sciences. He has a 40-year industrial background in analytical chemistry, polymers, and pharmaceuticals. In the past several years, he has been conducting research in the area of CEA and specifically vertical soil-less aeroponic growing technologies. He is currently also the R&D Director for Aero Development Corp, a maker of commercial aeroponic growing systems.

1 Introduction

Agriculture ... is our wisest pursuit, because it will in the end contribute most to real wealth, good morals, and happiness.

Thomas Jefferson

Aeroponics is a new emerging growing technology that is best defined as a soil-less method for growing plants in which their roots are suspended in air. Nutrients are provided to the roots either by misting a nutrient solution or by trickle down gravity flow of the nutrient solution. Hydroponics is a much more familiar technology to the public because it has been in practice for almost 100 years. Hydroponics is similar to aeroponics except that the roots of the plants are submerged in water, also called water cultures. The roots of these plants absorb nutrients from the nutrient solution that the roots are in constant contact with. These technologies will be further discussed in more detail in Chapter 3 of this book. There are many similarities between the development of aeroponics and the development of hydroponic systems. For example, the nutrient solutions, pH, electrical conductivity, and other parameters are very similar for these two sister technologies. There is also some overlap with what is called controlled-environment agriculture (CEA). Aeroponics and hydroponics are both considered to be CEA technologies because they are normally practiced in a controlled environment, such as a greenhouse, a warehouse, or a shipping container, where many of the environmental variables are controlled. This would include temperature, light intensity, photoperiod, nutrient concentration, humidity, carbon dioxide levels, etc. There can also be overlap with soil-based growing (geoponics) that is conducted in a greenhouse or warehouse.

Why is aeroponics important? It is important mainly because the future production of good, pure, and safe food is uncertain. In an article in *Newsweek*, the question was asked, how are we going to feed humankind in the future if we keep farming like we've been for the past century (*Newsweek*, 2015)? The current population on the earth is about 7 billion humans and by 2050 that number is projected to be nearly 10 billion. The problem is that most of the land we can use for food production is already being cultivated; which means that we are going to have to make some large-scale changes to how we farm.

PROFESSOR DESPOMMIER

Professor Dickson Despommier (Columbia University) tells the story of a Florida farmer who had a 30-acre strawberry farm that was destroyed in 1992 by Hurricane Andrew. The farmer obtained the insurance money to rebuild his farm, but instead of replanting strawberries, he used the money to build a greenhouse: "He did this because he thought if he built the greenhouse strong enough it might survive the next hurricane, and he was right," says Despommier. His hydroponic greenhouse was so efficient that 1 acre of indoor space could grow more strawberries than the farmer had previously been able to produce on 30 outdoor acres—30-fold increase. This left the farmer with 29 acres of unused land (*Newsweek*, 2015).

In his book *Vertical Farms* (2011), Despommier describes a vision for urban agriculture in what is being described now as CEA. He contrasts yesterday's agriculture with today's, and projects the future of agriculture evolving into vertical farms. He describes the advantages and all the benefits of this innovative idea. He proposes that the need for food will be where the people are in the cities so the food should be grown close to where they are. The economic benefit is obvious, that is, the reduction in shipping costs for food being shipped thousands of miles from farm to table. He summarizes the following four key themes that would be necessary to implement vertical farming:

1. capture sunlight and disperse it evenly among the crops;
2. capture passive energy for supplying a reliable source of electricity;
3. employ good barrier design for plant protection;
4. maximize the amount of space devoted to growing crops.

He shows several pictures of futuristic multistory buildings with crops growing on every floor. He concludes his book with a chapter titled "Food Fast-Forward" in which he concludes that disruptive technology is simple. It disrupts the present and jump-starts the future. The vertical farm has the potential to do that by advancing agriculture to a place in history it has never before occupied, which is one of true sustainability. His recommendation is a revamping of the United States Department of Agriculture (USDA) to help facilitate this transformation (Despommier, 2011).

Despommier in his TEDx Middleburg talk in 2013 listed ten benefits of vertical urban farming (Despommier, 2013):

1. no agricultural runoff;
2. year-round crop production;
3. no crop loss due to severe weather;
4. uses 70% less water, no agrochemicals, no fossil fuels;
5. allows for the restoration of the ecosystem;
6. remediates gray water;
7. creates jobs in the city;
8. supplies fresh produce to city dwellers;
9. uses abandoned city properties;
10. can grow biofuels and drugs.

He cites examples of vertical farms already in place around the world.
Examples are:

Rural Development Agency Suwon, Korea
Nuvege Kyoto, Japan
SkyGreen, Singapore
TerraSphere Vancouver, Canada

Plantagon Linkosing, Sweden
The Plant Chicago, USA
Vertical Harvest, Jackson Hole, USA

OTHER TED TALKS

Christine Zimmermann-Loessl, the head of the Association of Vertical Farming in Munich, Germany, gave a TED talk in Liege in 2015 entitled, "Taking food production to new heights."

She presented the "real" reasons for the benefits of vertical farming. She gave the main three reasons—fresh, safe food; less use of natural resources; and less traveled food miles and spoilage. She emphasized the need for increased food production based on the projected population growth from 7 to 10 billion people in the next 30 years. She presented the picture of vertical farming or CEA as a utopia for growing produce. Her presentation included the ideal conditions for optimum growth—temperature, light, water, nutrients, specific light spectrum, light duration, and protection from severe weather conditions. The benefits were presented as two to three times the growth rate, reduction in land use by 10-fold, and produce that is rich in vitamins, minerals, and antioxidants as well as tastes good. This she stated was one of the pillars of the future of agriculture (Zimmermann-Loessl, 2015).

According to a TED talk in Tanzania in 2017, Sara Menker indicated that by 2050, there will be a need for a 70% increase in worldwide food production. Can we feed the population using our current methods (Menker, 2017)? According to an article in the *Atlantic* magazine the world is divided into two groups—the wizards and the prophets. The wizards believe that we can technically innovate and solve this challenge. The prophets believe that we need to conserve, reduce, and adjust to this new reality or we won't survive (Mann, 2018). Achieving this goal will most likely take several technical solutions to address this issue. These include traditional farming methods and many new technologies.

NEW TECHNOLOGY

Companies like Aqua Design are capturing the CEA idea for the urban dweller. Toni Beck, their chief marketing officer says, "For many people who live in urban areas, like New York, it's rare that you have a backyard or even enough indoor space to grow your own healthy veggies so we designed EcoQube Sprout for the urban dweller who wants fresh greens but just doesn't have enough space or time. We believe that the future of food production is through the use of aquaponics, hydroponics, and aeroponics. With these technologies we can grow food 30%–50% faster while using 90% less water," said Beck. "We can grow more efficiently using less space and less water, allowing us to produce more food. It was really important for us to design the Sprout for everyone, of all ages, to grow their own fresh food easily indoors" (Mashable.com, 2018).

The Tabernas desert, in southern Spain, is the driest place in Europe. But in the 1960s the land began to blossom, and today the arid desert is where more than half of Europe's fresh vegetables and fruits are grown (Tremlett, 2005). The credit goes to greenhouses. The first few were built there in 1963, courtesy of a land distribution project spearheaded by Spain's Instituto Nacional de Colonización. Fruits and vegetables from those greenhouses, where the environment could be controlled and beautiful produce could be grown, consistently soon outsold comparable crops grown elsewhere in open fields. Money was reinvested, greenhouses were expanded—with inexpensive plastic sheeting replacing glass as the material of choice for the majority of the controlled environments—and today greenhouses cover 50,000 acres in the Tabernas desert, adding $1.5 billion annually to the economy of Spain.

That's because from an environmental and land-use perspective, controlled-environment farming is a great idea. Fruits and vegetables grown indoors tend to have far greater yields per area than comparable produce grown outside. Put a roof and walls around produce, and thus, most problems caused by weeds, pests, and inclement weather vanish. Add technology like hydroponics—growing plants so the roots sit in a customized nutrient slurry instead of in plain old dirt—to the equation, and yields increase even more. Better yet, build a hydroponic rig that is modular, rotates, and stacks—which means you can have several "stories" of produce growing atop the same ground (assuming the stacks all get sufficient light).

In 2011, a calamity in Japan made it necessary to rethink agricultural production strategies. The tidal wave that caused the Fukushima disaster wiped out most of the farmland near Sendai, a coastal area in the northern half of Honshu, the largest island of Japan. The Japanese government decided to jump-start a vertical farm building boom, there in an effort to replace the lost land. Four years later, Japan boasts hundreds of vertical farms, greenhouses stacked high into multistory skyscrapers, where plants rotate daily to catch sunlight. Instead of transporting dirt into the buildings, the plants grow with roots exposed, soaking in nutrients from enriched water or mist.

The number of Japanese plant factories (PFs) producing more than 10,000 heads of lettuce daily is estimated to be around ten. Japan's PFs are expanding to meet the increasing demand for safe, pesticide-free, and locally grown food. Japan has more PFs than any other country. The largest number of PFs are located in Okinawa Prefecture near Taiwan. The rapid commercialization and financial subsidization by the Japanese government of PFs, which began in 2010, are helping to drive interest in their development (Kuack, 2017).

Another reason for the increase in PFs in Japan is that the country has been importing a large amount of fresh, sliced salad vegetables from China. The Japanese are concerned about the amount of pesticides being used for Chinese vegetable production and looking for alternative sources of fresh vegetables and herbs.

In 2014, there were about 170 PFs in Japan. Of these, 70 are producing more than 1,000 lettuce heads (50–100 g per head) or other leafy greens daily. The average floor area of a PF with 10–15 tiers for producing 10,000 lettuce heads daily is 1,500 square meters. The main components of a PF are:

1. a thermally well-insulated and airtight warehouse-like structure with no windows;
2. tiers/shelves with a light source and culture beds;
3. a carbon dioxide supply unit;

4. nutrient supply units;
5. air conditioners;
6. an environment control unit;
7. other equipment includes nutrient solution sterilization units, air circulation units, and seeders.

Aeroponics, a companion technology to hydroponics, has taken off in Japan and is helping high-tech greenhouses produce remarkable yields quickly: unlike hydroponic systems, where plants dip their roots in nutrient slurry, aeroponic systems spray the plants' deliberately exposed roots with a nutrient-laden mist. "The root systems grow much longer because they have to increase their surface area to absorb the same amount of nutrients," explains Despommier (Kozai, 2016b). That, in turn, makes the plants grow much faster.

VERTICAL FARMING TRENDS

Singapore, Sweden, South Korea, Canada, China, and the Netherlands all now boast skyscraper farms similar in concept to Japan's. In the US, such farms have risen in Chicago, Newark, New Jersey, and Jackson, Wyoming.

In the UK and the Netherlands, in Boston and in Bryan, Texas, it's been done. "Pinkhouses," as they're sometimes called, are lit blue and red: those are the spectrums of visible light best absorbed by plants. By using these colors alone, pinkhouses generate serious efficiency. In the wild, plants use at most 8% of the light they absorb, while in pinkhouses, the plants can use as much as 15%. In addition, because everything happens entirely indoors, the lights, temperature, and humidity can be controlled to an extent not possible even in the most high-tech, sun-dependent vertical farms and greenhouses (Kozai,T. 2016a).

As a result, the plants grown in these pinkhouses grow 20% faster than their out-door cousins, and need 91% less water, negligible fertilizer, and no treatment with herbicides or pesticides. Currently, the LEDs keep the upfront costs of constructing a pinkhouse very high, but LED prices are projected to drop by half in the next five years. Given that, perhaps we ought to be preparing for a future where the majority of our produce is grown industrially in LED-lined skyscrapers made of steel and poured concrete.

AEROPONICS AT NASA

Plants have been to space since 1960, but NASA's plant growth experiments began in earnest during the 1990s. Experiments aboard the space shuttle and International Space Station have exposed plants to the effects of microgravity. These experiments use the principles of aeroponics: growing plants in an air/mist environment with no soil and very little water.

In the 1990s and 2000s, NASA conducted research on aeroponic growing of food for space applications partnering with AgriHouse and BioServe Space Technologies. This technology was targeted for a microgravity environment on the Mir space station. The objective was to produce plants free of infection without using pesticides.

Richard Stoner II, president and founder of AgriHouse, began using aeroponics in the late 1980s to grow herbs in a greenhouse. Utilizing his own patented aeroponic process, Stoner was one of the only people in the US employing the aeroponic plant-propagating technique at the time.

The adzuki bean seeds and seedlings sprouted quite well both on Earth and aboard the space station. The Mir-grown seeds and seedlings, however, exhibited more growth than those grown on Earth. These plants have developed healthy root systems, all while growing in the soil-less environment of the Genesis Series V aeroponic V-shaped rapid growth system. NASA developed a low-mass, inflatable aeroponic system for rapid crop production of pesticide-free herbs, grains, tomatoes, lettuce, peppers, and other vegetables. This clean, sterile environment greatly reduces the chances of spreading plant disease and infection that is commonly associated with other growing media. Each growing chamber has 161 openings. The grower can place one to five cuttings in each opening. Plants do not stretch or wilt while they are developing their roots. Once roots are developed, the plants can be easily removed for transplanting into any type of media without experiencing transplant shock or setback to normal growth. Despite the drastic reduction in water and fertilizer usage, those employing the aeroponics growing method witnessed robust crop yields and healthy, vibrant coloring.

The Genesis system can grow many different plant types, including nursery stock, fruits, vegetables, and houseplants. Hundreds of varieties have been cultivated by researchers, farmers, commercial greenhouse growers, and nursery operators. In the case of tomatoes, for example, growers can utilize the soil-less method to get a jump-start on their production. Tomato growers traditionally start their plants in pots, having to wait at least 28 days before transplanting them into the ground. With the Genesis system, growers can start the plants in the aeroponic growing chamber, then transplant them to another medium just 10 days later. This advanced aeroponic propagation technology offers tomato growers six crop cycles per year, instead of the traditional one to two crop cycles.

According to AgriHouse, growers choosing to employ the aeroponics method can reduce water usage by 98%, fertilizer usage by 60%, and pesticide usage by 100%, all while maximizing their crop yields by 45%–75%. By conserving water and eliminating harmful pesticides and fertilizers used in soil, growers are doing their part to protect the Earth. These results essentially proved that aeroponically grown plants uptake more minerals and vitamins as compared to other growing techniques.

According to AgriHouse, potato production in East Asia lags behind North America due to poor performance of seed potato crops. Utilizing the closed-loop features developed under the NASA grants, the company designed and installed a state-of-the-art aeroponic potato laboratory at the Institute for Agrobiology, for potato tuber seed production. "AgriHouse's advanced technology gives the Institute of Agrobiology the opportunity for a direct replacement of labor-intensive, *in vitro* tissue culture potato production," said Dr. Nguyen Quang Thach, the institute's director. "Furthermore, the economic impact in the region from the seed potato-production features of this NASA technology will give our underdeveloped country a tremendous boost."

The Flex system, however, possesses a chamber that contains 1,000 plant holders, offering a 10-fold increase in fresh crop production per square meter over the Genesis system. It is capable of delivering 12 growing cycles per year and eliminates the need for a greenhouse.

What NASA has learned from their research is that aeroponic growing systems provide clean, efficient, and rapid food production. Crops can be planted and harvested in the system all year round without interruption, and without contamination from soil, pesticides, and residue. Since the growing environment is clean and sterile, it greatly reduces the chances of spreading plant diseases and infection commonly found in soil (Spinoff, 2006).

AEROPONICS AT DISNEYWORLD

Tim Blank envisioned the future of growing while working at Disneyworld in the 1990s in the Disney Park called the Land. There he did research and developed the aeroponic technology. In 2004, he launched a company called Tower Gardens and since then has sold thousands of patented vertical aeroponic Tower Gardens around the world. He calls it the power of the tower. When he is asked "What is aeroponics?" He replies, "Aeroponics is simply defined as the process of growing plants in an air or mist environment without the use of soil or an aggregate medium. The Tower Garden® growing chamber contains no soil or aggregate medium. Instead, the chamber is empty. It's just roots and air between each irrigation cycle. The tumbling water during these irrigation cycles creates a fine mist, oxygenating the water and bathing the roots of each plant on its way down to the reservoir. This process is continuously repeated with each irrigation cycle, providing maximum amounts of fresh oxygen, water, and nutrients to the roots of the plants 24 hours a day. The intelligent design of the Tower Garden® system produces extraordinary crops that grow much faster than they would in soil, producing bountiful harvests within weeks of being transplanted into the system."

One of the main purposes behind Future Growing®'s (formerly Tower Garden) patented aeroponic design was to avoid clogging misters—which typically plagues traditional aeroponic growing systems—by utilizing high-flow aeroponics. Another key benefit is the massive growing chamber for the roots. Because the plants' roots do not run out of space, they continue to grow strong and healthy. There are commercial Tower Garden® farmers producing herb crops for several years now with plenty of room to go!

To achieve their mission of producing healthy food for people, they also developed an all-natural, stable, water-based ionic mineral solution to support the patented vertical aeroponic Tower Garden® technology.

With assistance from leading world experts in plant and human nutrition, they developed the proprietary Aeroponic Power-Gro® and the Tower Tonic® plant food. Aeroponic Power-Gro® and the Tower Tonic® contains a wide range of specially formulated ionic minerals and plant nutrients. It is the world's first high-performance ionic mineral solution specifically designed for all types of food and flowering crops. The pH-balanced blend of natural plant nutrients helps stimulate plants' roots, flowers, fruits, and leaves.

Unlike conventional hydroponic fertilizers, the amazing Aeroponic Power-Gro® and the Tower Tonic® can be used to grow everything from gourmet lettuce and edible flowers to beautiful vine-ripened tomatoes. Healthy plants packed with nutrition help create healthy people.

Aeroponic Power-Gro® and the Tower Tonic® are also loaded with trace minerals that are essential to vibrant human health! Jake Kelly, a commercial rooftop Tower Garden® farmer in Southern California, recently grew a 3-foot aeroponic kale plant in a matter of weeks (Blank, 2020). Several more stories of home and commercial use of these systems is given in Chapter 3—the Business of Aeroponics.

UGLY FOOD

Aeroponic growing can also reduce the amount of food lost due to its appearance.

In the US, as much as 40% of produce grown is never sold or eaten. The reason? It's too ugly. Consumers won't buy imperfect looking fruit and vegetables, and grocery stores refuse to stock them. The demand for "pretty" produce means fruit and vegetable farmers need to make up for the cost of all that ugly food they can't sell.

That's also why controlled environments, from pinkhouses in Boston to plastic-sheeted greenhouses in Almeria, are used overwhelmingly to grow fresh produce: farmers who work in controlled environments can put out consistently pretty pieces of produce. They have a huge advantage in the current fruit and vegetable market, which values the look of the crop as much as anything. Moreover, with produce, freshness fetches a premium; the shorter a distance a piece of produce has to travel before it reaches your plate, the tastier it'll be and the more you'll pay for it. And controlled environments allow farmers to grow their produce right next door to where it's sold. That's why, even in the land-rich US, says Chieri Kubota, a professor at the University of Arizona's School of Plant Sciences, 40% of tomatoes today sold fresh in stores are grown in greenhouses.

However, controlled-environment farming is far less profitable for growers of staples. Rice, corn, and wheat—the cereal grains that provide the world with about 50% of its calories—are all dirt-cheap, more or less regardless of appearance. The margins on those crops are thin, so any additional investment in innovation and production methods comes at an impossibly steep price. Staple farmers can see their profits only by growing huge amounts of their crops on enormous swaths of land; economically, it doesn't make sense for them to try to replicate that profit model in greenhouses, so controlled-environment farming is unlikely to supplant the open field when it comes to our most important crops (*Newsweek*, 2019).

VERTICAL FARMING DEFINITION

The term "vertical farm" may be a bit confusing. There are several ways to grow produce vertically. Hydroponics—growing produce with the roots immersed in water—has been the main technique for soil-less agriculture for the past 100 years. However, there are several methods of growing hydroponically. In Jone's book, *Hydroponics—A Practical Guide to the Soil-less Grower*—several methods are described (Phillips, 2019). They are divided into two groups—medium-less hydroponics (which includes aeroponics) and medium hydroponics. Medium being a solid substrate that the roots can attach to. The medium-less also includes the standing aerated nutrient solution and nutrient film technique. The medium systems include ebb and flow, drip/pass through inorganic medium systems, bags/buckets, or rock wool slabs.

In the case of vertical farms, they can be based on horizontal growing trays or vertical walls or vertical columns. Normally the horizontal trays would be based on one of the above-mentioned hydroponic systems with the trays stacked vertically. They require artificial light between trays to ensure adequate radiation energy for the plants to grow. Technically, this method is hydroponics. On the other hand, vertical walls or columns are normally called aeroponic because the roots are suspended vertically in air and the roots are either misted with nutrient solution or dosed with a stream of nutrient solution.

AEROPONIC CROPS

While a wide variety of fruits, vegetables, and edible/medicinal plants can be grown for commercial production, space constraints continue to limit those that make the most financial sense. The most common crops grown for commercial production are lettuces, salad greens, and culinary herbs. Recent research from the Cornell University Cooperative Extension has shown that hydro/aeroponics is the most efficient method for growing leafy greens. Leafy greens grown using traditional geoponic agriculture can become contaminated with bacteria and soil pathogens. The Cornell research shows that hydro/aeroponics significantly reduce these risks.

To provide a better understanding of which plants can be grown using hydro/aeroponics, here is a partial list of crops that have been grown successfully:

1. arugula;
2. basil, all varieties;
3. beans;
4. bok choi;
5. brussels sprouts;
6. cabbage;
7. chard;
8. cress;
9. cucumbers;
10. culinary herbs, including cilantro, mint, oregano, thyme, dill, rosemary;
11. fennel;
12. flowers including nasturtium, violas, marigolds, poppies, lavender;
13. kale;
14. lettuces and salad greens;
15. medicinal herbs;
16. microgreens;
17. mustard greens;

18. okra;
19. ornamental plants;
20. pea
21. spinach
22. strawberries
23. sweet and hot peppers
24. all tomato varieties.

It has been reported that root vegetables and below-the-soil crops are more difficult to grow with hydro/aeroponics. There are research groups, however, such as at the Ha Noi University of Agriculture, who are working to develop seed potato crops (Brechner and Both, 2013).

AEROPONIC CONTAINER FARMS

One of the newest variations on aeroponic technologies is the farm in a shipping container concept. An entire aeroponics farm can be placed inside of a former cargo shipping container. An applied spray is concentrated with essential macro- and micronutrients, typically provided from purchased chemicals, however, not requiring pesticides. Container farmers have the ability to control the entire growing environment (another form of CEA) by keeping the plants in an enclosed space and climate-controlled—a farm in a box.

Companies like Vertical Roots, Modular Farms, and Freight Farms are leading the way with smart data that allows for automated adjustments to container temperature, light intensity, carbon dioxide levels, and nutrient concentration. These boxes have several environmental and business benefits as well. Vertical Roots, for example, is able to mitigate 95% of its water loss by simply catching the runoff and circulating it back to the system's reservoir. All of the growing operations occur within a shipping container box, a 40 × 8 × 8-foot space. Its plants are suspended vertically, meaning their container model can optimize space to grow produce normally requiring 4 acres of land within a 1–2-month schedule at a fraction of the acreage of a traditional farm. This design concept, also used by Freight Farms, not only conserves space but also allows for the farms to be portable to different locations.

Information provided by Vertical Roots shows that their pod method, which includes four boxes of farms, can produce roughly $200,000 of yearly income on leafy greens. An expensive purchase price of over $500,000 would likely keep any single person from buying the pod themselves, but utilizing municipal support funds, loan programs, and philanthropic aid could offset the daunting first step. Upon running operations with no debt or loan paybacks, a single pod could provide well-paying jobs for five or more people.

Every system has its drawbacks, and aeroponics is no exception. The systems are typically located in an enclosed space for climate-control optimization, which requires high-intensity light systems, temperature controls, and other monitoring equipment and this comes with an expensive monthly electric bill. Environmentalists also note that producing the shipping container, ordering the container to be shipped to your location, stripping it of its potential toxins, and outfitting it with the advanced technology greatly diminishes the sustainability of the system (Miller, 2018).

These concerns are valid, and growing methods should be improved upon to reduce the carbon footprint. Nonetheless, aeroponics farming concepts may present a cleaner, more community-driven alternative when compared to the traditional farming methods that take up hundreds of acres of land, several states away. As our cities expand and poor communities find themselves with less access to quality food, organizations can lead the way in bringing food into the neighborhoods that need it most. Successful business models that capitalize on aeroponics' ability to optimize space and resources can not only produce fresh food but also reduce our farming carbon footprint.

The next chapter will look at the history of aeroponics and some of the key technical developments that were the foundation for soil-less agriculture. This will include the origin of the terms—hydroponics and aeroponics—and some of the early technical papers about aeroponics. In subsequent chapters, a detailed summary of the most current research and innovation developments that have usher in this new era of aeroponic technology will be discussed.

REFERENCES

Blank, T., *The Power of the Tower*, futuregrowing.wordpress.com – Accessed April 16, 2020.

Brechner, M., Both, A. J., 2013, *Cornell Lettuce Handbook*. cea.cals.cornell.edu/attachments/Cornell%20CEA%20Lettuce%20Handbook%20.pdf.

Despommier, D., 2011, *The Vertical Farm: Feeding the World in the 21st Century*. London: Picador.

Despommier, D., 2013, TEDx talk Middlebury. www.youtu.be/XO2mVBTeBtE.

Kozai, T., 2016a, *LED Lighting for Urban Agriculture*. Singapore: Springer.

Kozai, T., 2016b, *Plant Factory: An Indoor Vertical Farming System for Efficient Quality Food Production*. Amsterdam: Elsevier.

Kuack, D., 2017, Japan plant factories are providing a safe, reliable food source, *UrbanAgNews.com*, May 15, 2017.

Mann, C., 2018, Can planet earth feed 10 billion people? – *The Atlantic*, www.theatlantic.com/magazine/archive/2018/03/charles-mann.../550928.

Mashable.com, 2018, Grow microgreens at home with Kickstarter campaign the EcoQube Sprout, *Mashable.com* – July 23, 2018.

Menker, S., 2017, TEDx talk Tanzania. www.ted.com/talks/sara_menker_a_global_food_crisis_may_be_less_than_a_decade_away/footnotes?language=en.

Miller, Matthew, 2018, Aeroponics: A Sustainable Solution for Urban Agriculture, April 4, 2018, www.eli.org/vibrant...blog/aeroponics-sustainable-solution-urban-agriculture.

Newsweek, 2015, To Feed Humankind, We Need the Farms of the Future Today – If we keep farming like we've been for the past century, we'll end up with millions starving and a planet denuded of trees. *Newsweek* – October 30, 2015.

Newsweek, 2019, www.newsweek.com/vertical-farms-across-world-385696, July 15, 2019.

Phillips, S., 2019, www.uglyproduceisbeautiful.com/ugly-produce-problem.html.

Spinoff, 2006, www.nasa.gov/vision/earth/technologies/aeroponic_plants.html.

Tremlett, G., 2005, Spain's greenhouse effect: The shimmering sea of polythene consuming the land, *The Guardian*, September 2005.

Zimmermann-Loessl, C., 2015, TEDx talk Liege. www.youtu.be/ecLMTgAWsqs.

2 History of Aeroponics

Yesterday is history, tomorrow is a mystery, today is God's gift, that's why we call it the present.

Joan Rivers

The history of aeroponics finds its roots in various examples of agricultural experiments and the development of technology (chemistry, biology, and engineering), especially materials, e.g., glass-making, the invention of rubber hoses and plastic, as well as steam and electricity (for lights and pumps). Its history is connected with the development of soil-less and hydroponic growing (Steiner, 1985).

In ancient times (600 B.C. to 300 A.D.), the "Hanging Gardens of Babylon" was one of the seven wonders of the world. Possibly one of the first examples of protected agriculture. These gardens were built by King Nebuchadnezzar II on the east bank of the Euphrates River in the middle of the desert for one of his wives. Renditions suggest a series of terraced growing areas in which water is supplied by a "chain pump" lift system from the river below. Egyptian hieroglyphs tell of the people growing plants in water culture, possibly papyrus (for paper) and lotus (University of Chicago, 1993). Theophrastus (372–287 B.C.) was one of the greatest early Greek philosophers and called the "father of botany." He performed experiments in crop nutrition and noted that rotting manure (compost) warms and ripens the soil, thus increasing growth; worked with potted plants (Grene and Depew, 2004). Cucumbers were grown off-season for the Roman Emperor Tiberius (14–37 A.D.) using a structure covered with "transparent rock" (presumably mica). It was the first known use of controlled environment agriculture (Janick, Paris, and Parrish, 2007). Other such structures were described during 1st century. Pliny "The Elder" aka Gaius Plinius Secundus (23–79 A.D.) wrote *Naturalis Historia*, a series of 37 books. Books 12–27 covered botany, agriculture, horticulture, and pharmacology (Healy, 2004). He talked about the use of "straw caps" (mulch) to protect young plants. In 300 A.D. in Rome, roses were forced to flower early by the addition of warm water into the irrigation ditches twice a day. This would warm the roots and stimulate growth. Therefore, up to ~300 A.D., the ancients had perfected protected agriculture (terraced growing areas, mulches, and compost heating), greenhouses, hot air and hot water heating systems, and had experimented with plant nutrition, water culture, and more. Then the Great Library burned in Alexandria, Egypt. Rome fell and the Dark Ages began (Rohrbaugh, 2015).

Early forms of hydroponic growing were first observed in China in the 1200s by Marco Polo, who traveled with his father and uncle along the "Silk Road" to China, and saw "floating gardens" used to grow food. This type of garden had presumably been used for centuries by the Chinese. The Aztecs in the 1300s, built Tenochtitlan on an island in the shallow lake, Texcoco (near present day Mexico City), and created artificial islands (floating gardens) called "chinampas." This was necessary since there was little dry/level land for farming.

In 1300–1500s the European Renaissance triggered a revival of art, literature & learning in the world as it emerged from the Dark Ages. People want to "grow out of season" (i.e., have tomatoes in winter) or grow plants where they don't normally grow. In 1385, the French built "glass pavilions" oriented toward the south to grow flowers (though mainly for the wealthy to enjoy) (Royal Netherlands Academy of Arts and Sciences, 2015). In the 1500s and 1600s, the expanding glass industry in Italy prompted Italians, English, Germans, and French to experiment with glass for plant growing. Greenhouses were built to grow flowers in the winter using solar heating, the building of "orangeries" for growing oranges and other citrus (to help guard against scurvy during the winter) and the first greenhouse or conservatory to be built at a botanical garden was constructed at Padua, Italy, in 1550 (Aeroponics: Wikipedia, 2019). In the 1600s, widespread use of glass greenhouses of different designs (including those with removable tops for summer) to "force" bulbs and to grow flowers, citrus, and other trees and shrubs in England, France, Germany, Sweden, The Netherlands, Spain, and China (Hoagland Solution: Wikipedia, 2014). Heating systems were "rediscovered"—steam/hot water, "bark" stoves (moist heat), manure as a heat source and charcoal heaters.

In 1600, a Belgian, Jan Van Helmont, performed the earliest known experiments to determine the constituents of plants. A 5 lb willow shoot planted in 200 lbs of soil was covered to keep dust out and watered with rain water for 5 years. The willow increased its weight to 160 lbs., but the soil lost only 2 oz. His conclusion: plants obtain substances from the water needed for growth [these "substances" were "elements" (not yet known)]. However, he failed to realize that plants also require carbon dioxide and oxygen from air. In 1699, an Englishman, John Woodward, used various types of soil to grow plants. He found that the greatest growth occurred in water which contained the most soil. His conclusion: plant growth results from substances in the water derived from the soil, rather than from the water itself. As with Van Helmont, the elements were not yet fully known.

In the 1700s, greenhouse designs continued to improve in Europe and then in the USA, including multispan structures. The first greenhouses with glass on all sides were built in 1700s (Carter, 1942). In the 1700s, European chemists discovered a majority of the elements (except carbon, sulfur, iron, and copper, which were discovered in ancient times) including those elements necessary for plant growth. Growers in The Netherlands found that glass cleaning along with greenhouse orientation (perpendicular to radiation source) are important for light penetration, especially in northern latitudes. George Washington built a glass conservatory with below-ground heating at his home at Mount Vernon in the 1780s. In the 1800s in the USA, the first commercial greenhouse (1820) was built.

In 1804, N.T. de Saussure made the first quantitative measurements of photosynthesis and proposed that plants are composed of chemical elements from soil, water, and air. Curiously enough, the earliest recorded experiment with water cultures (soilless growing) was carried out in search of a so-called "principle of vegetation" in a day when so little was known about the principles of plant nutrition that there was little chance of profitable results from such an experiment. Woodward in 1699 grew spearmint in several kinds of water: rain, river, and conduit water to which he added garden mold in one case. He found that the greatest increase in the weight of the plant took place in the water containing the greatest admixture of soil. His conclusion was "That earth, and not water, is the matter that constitutes vegetables."

The real development of the technique of water culture took place in the 19th century. It came as a logical result of the modern concepts of plant nutrition. By the middle of the 19th century, enough of the fundamental facts of plant physiology had been accumulated and properly evaluated to enable the botanists and chemists of that period to correctly assign to the soil the part which it plays in the nutrition of plants. They realized that plants are made of chemical elements obtained from three sources: air, water, and soil; and that the plants grow and increase in size and weight by combining these elements into various plant substances.

Water of course, always the main component of growing plants. But the major portion, usually about 90%, of the dry matter of most plants made up of three chemical elements: carbon, oxygen, and hydrogen. Carbon comes from air, oxygen from air and water, and hydrogen from water. In addition to the three elements named above, plants contain other elements, such as nitrogen, phosphorous, potassium, and calcium, which they obtain from the soil. The soil, then, supplies to the plant a large number of chemical elements, but they constitute only very small portion of the plant. However, various elements which occur in plants in comparatively small amounts are just as essential to growth as those which compose the bulk of plant tissues.

The publication, in 1840, of Liebig's book on the application of organic chemistry to agriculture and physiology, in which the above views were ably and effectively brought to the attention of plant physiologists and chemists of that period, served as a great stimulus for the undertaking of experimental work in plant nutrition. (Liebig, however, failed to understand the role of soil as the source of nitrogen for plants, and the fixation of atmospheric nitrogen by nodule organisms was not known then.)

Once it was recognized that the function of the soil in the economy of the plant to furnish certain chemical elements, as well as water, was but natural to attempt to supply these elements and water independently of soil. The credit for initiating exact experimentation in this field belongs to the French chemist, Jean Boussignault, who was known as the founder of modern methods of conducting experiments in vegetation.

Boussignault, who had begun his experiments on plants even before 1840, grew them in insoluble artificial soils: sand, quartz, and sugar charcoal, which he watered with solutions of known composition. His results provided experimental verification for the mineral theory of plant nutrition as put forward by Liebig, and were at once a demonstration of the feasibility of growing plants in a medium other than a "natural soil."

In 1851, the French chemist, Jean Boussingault, verified de Saussure's proposal when he grew plants in insoluble artificial media (sand, quartz, and sugar charcoal) and solutions of known chemical composition (Stoner, 1983). His conclusions: plants require water and obtain hydrogen from it; plant dry matter contains hydrogen and carbon and oxygen which comes from the air; plants also contain nitrogen and other elements.

This method of growing plants in artificial insoluble soils was later improved by Salm-Horstmar (1856–1860) and has been used since, with various technical improvements, by numerous investigators throughout the world. In recent years, large-scale techniques have been devised for growing plants for experimental or commercial purposes in beds of sand or other inert solid material. After plants were successfully grown in artificial culture media, it was but one more step to dispense with any solid medium and attempt to grow plants in water to which the chemical elements required by plants were added.

One of the key scientists to work on understanding photosynthesis and how plants grow was Julius van Sachs. When he was 16 years old, his father died, and in the next year both his mother and a brother died of the cholera. Suddenly without financial support, he was fortunate to be taken into the family of Jan Evangelista Purkyně who had accepted a professorship at the University of Prague. Sachs was admitted to the university in 1851. Sachs famously labored long hours in the laboratory for Purkyně, and then long hours for himself each day after his work in the laboratory was finished. After the laboratory, he could devote himself entirely to establishing how plants grow (Carter, 1942).

In 1856, Sachs graduated with a doctor of philosophy, and then adopted a botanical career, establishing himself as Privatdozent for plant physiology. In 1868, he accepted the chair of botany in the University of Würzburg, which he continued to occupy (in spite of calls from more prestigious German universities) until his death.

Sachs achieved distinction as an investigator, a writer, and a teacher; his name will ever be especially associated with the great development of plant physiology which marked the latter half of the 19th century, though there is scarcely a branch of botany to which he did not materially contribute. His earlier papers, scattered through the volumes of botanical journals and of the publications of learned societies (a collected edition was published in 1892–1893), are of great and varied interest. Prominent among them is the series of "Keimungsgeschichten," which laid the foundation of our knowledge of microchemical methods, and also of the morphological and physiological details of germination. Then there is his resuscitation of the method of "water-culture," and the application of it to the investigation of the problems of nutrition. Most important are his experiments, developing the concept of photosynthesis, that the starch-grains, found in leaf chloroplasts, depend on sunlight. A leaf that has been in sunlight, then bleached white and stained with iodine turns black, proving its starch content, whereas a leaf from the same plant that has been out of the sun will remain white. Julius von Sachs collaborated with Wilhelm Knop in the 1860s to use "nutri-culture" (today called water culture, a type of hydroponics) to study plant nutrition. Plant roots were immersed in water that contained "salts" of nitrogen (N), phosphorus (P), potassium (K), magnesium (Mg), sulfur (S), and calcium (Ca). They found that these elements were needed in large amounts by the plant, hence the term "macronutrients" was given. Both scientists also devised nutrient solution recipes.

Over the course of the following 80 years, several other scientists studied plant mineral nutrition using water culture (hydroponics) and identified other elements needed by plants in much smaller amounts. These are called "micronutrients" and include iron (Fe), chlorine (Cl), manganese (Mn), boron (B), zinc (Zn), copper (Cu), and molybdenum (Mo) (Carter, 1942).

The work by Boussingault and de Saussure was confirmed in 1860 by Sachs and about the same time by Knop. To quote Sachs directly:

> In the year 1860, I published the results of experiments demonstrated that land plants are capable of absorbing their nutritive matters out of watery solutions, without the aid of soil, and that it is possible in this way not only to maintain plants alive and growing for a long time, as had long been known but also to bring about a vigorous increase of their organic substance and even the production of seed capable of germination.

The original technique developed by Sachs for growing plants in nutrient solutions is still widely used, essentially unaltered. He germinated the seed in well-washed sawdust, until the plants reached a size convenient for transplanting. After carefully removing and washing the seedling, he fastened it into a perforated cork, with the roots dipping into the solution. Since the publication of Sachs's standard solution formula (Table 2.1) for growing plants in water culture, many other formulas have been suggested and widely used with success by many investigators in different countries. Knop, who undertook water-culture experiments at the same time as Sachs, proposed in 1865 a nutrient solution, which became one of the most widely employed in studies of plant nutrition.

Other formulas for nutrient solutions have been proposed by Tollens in 1882, by Schimper in 1890, by Pfeffer in 1900, by Crone in 1902, by Tottingham in 1914, by Shive in 1915, by Hoagland in 1920, and many others. At the very inception of the water-culture work, investigators clearly recognized that there can be no one composition of a nutrient solution which is always superior. Thus, Sachs wrote:

> I mention the quantities (of chemicals) I am accustomed to use generally in water cultures, with the remark, however, that a somewhat wide margin may be permitted with respect to the quantities of the individual salts and the concentration of the whole solution—it does not matter if a little more or less of the one or the other salt is taken— if only the nutritive mixture is kept within certain limits as to quality and quantity, which are established by experience.

TABLE 2.1
Standard Solution Formulas for Water Culture

Sach's Solution 1860		Knop's Solution 1865		Pfeffer's Solution 1900		Crone's Solution 1902	
Ingredient	g/L	Ingredient	g/L	Ingredient	g/L	Ingredient	g/L
KNO_3	1.00	$Ca(NO_3)_2$	0.8	$Ca(NO_3)_2$	0.8	KNO_3	1.00
$Ca_3(PO_4)_2$	0.50	KNO_3	0.2	KNO_3	0.2	$Ca_3(PO_4)_2$	0.25
$MgSO_4$	0.50	KH_2PO_4	0.2	$MgSO_4$	0.2	$MgSO_4$	0.25
$CaSO_4$	0.50	$MgSO_4$	0.2	KH_2PO_4	0.2	$CaSO_4$	0.25
NaCl	0.25	$FePO_4$	trace	KCl	0.2	$FePO_4$	0.25
$FeSO_4$	trace			$FeCl_2$	trace		

Until recently, the water-culture technique was employed exclusively in small-scale, controlled laboratory experiments intended to solve fundamental problems of plant nutrition and physiology. These experiments have led to the determination of the list of chemical elements essential for plant life. They have thus profoundly influenced the practice of soil management and fertilization for purposes of crop production. In recent years, great refinements in water-culture technique have made possible the discovery of several new essential elements. These, although required by plants in exceedingly small amounts, often are of definite practical importance in agricultural practice. The elements derived from the nutrient medium that are now considered to be indispensable for the growth of higher green plants are nitrogen, phosphorous, potassium, sulfur, calcium, magnesium, iron, boron, manganese, copper, and zinc. New evidence suggests that molybdenum may have to be added to the list. Present indications are that further refinements of technique may lead to the discovery of additional elements, essential in minute quantities for growth.

In addition to the list of essential elements, which is obviously of first importance in making artificial culture media for growing plants, a large amount of information has been amassed on the desirable proportions and concentrations of the essential elements, and on such physical and chemical properties of various culture solutions as acidity, alkalinity, and osmotic characteristics. A most important recent development in water-culture technique has been the recognition of the importance of special aeration of the nutrient solution for many plants, to supplement the oxygen supply normally entering the solution when in contact with the surrounding atmosphere.

The recently publicized use of the water-culture technique for commercial crop production does not rest on any newly discovered principles of plant nutrition other than those discussed above. It involves rather, the application of a large-scale technique, developed on the basis of an understanding of plant nutrition gained in previous investigations conducted on a laboratory scale. The latter have provided knowledge of the composition of suitable culture solutions. Furthermore, methods of controlling the concentration of nutrients and the degree of acidity are, except for modifications imposed by the large scale of operations, similar to those employed in small-scale laboratory experiments (Zobel, Del Tredici, and Torrey, 1976; Rohrbaugh, 2015).

In 1911, Vladimir M. Artsikhovski, a Russian botanist, published in the journal *Experienced Agronomy* an article *On Air Plant Cultures* which describes his method of studies on root systems by spraying various substances in the surrounding air—the aeroponics method (Rohrbaugh, 2015).

In the first half of the 20th century, several plant nutrition scientists developed recipes for optimum plant growth. Several researchers in the 1930s began modifying small-scale laboratory techniques for liquid culture (hydroponics) to accommodate large-scale commercial crop production. In 1929, W.F. Gericke (U.C. Berkley) introduced "soil-less" culture and grew tomato plants in nutriculture and experimented with this on a large scale for commercial purposes. He coined the term "hydroponics" in 1937, from two Greek words "hydro" meaning "water" and "ponos" meaning "work." Literally = "water working." However, Gericke met up with skepticism from the public and the university. His colleagues even denied the use of the on-ground greenhouses for his study. He declared them wrong by successfully growing 25-foot tall tomato plants in nutrient-filled solutions. The university still doubted his account of successful cultivation and the Director of the California Agricultural Experimental Station, C. B. Hutchinson, requested two other students, Dennis R. Hoagland and Daniel I. Arnon, to investigate his claim. The two performed the research and reported their findings in an agriculture bulletin 1938, titled "The Water Culture Method for Growing Plants without Soil" (Hoagland and Arnon, 1938). They confirmed the application of hydroponics but concluded from their research that crops grown with hydroponics are no better than those grown on quality soils. However, they missed many advantages of hydroponics in comparison with the cultural practice. Crop yields were ultimately limited by factors other than mineral nutrients, especially light. This research, however, overlooked the fact that hydroponics has other advantages including the fact that the roots of the plant have constant access to oxygen and that the plants have access to as much or as little water as they need. Hoagland became so well known for his work in plant nutrient formulas that today it is common to refer to his hydroponic nutrient solution recipe as a "MODIFIED HOAGLAND'S SOLUTION." The Hoagland solution is shown in Table 2.2.

In the 1940s, during World War II, the United States military used hydroponics to supply the troops stationed on isolated, nonarable islands and coral atolls in the Pacific with food. After the war, the U.S. Army built a 22 hectare hydroponic operation at Chofu, Japan. In the 1940s, the technology was largely used as a research tool rather than the economically feasible method of crop production. W. Carter in 1942 was the first grow plants in air called air culture growing and described a method for growing plants in water vapor to facilitate the study of roots (Julius von Sachs, 2019).

TABLE 2.2

Hoagland Solution for Hydroponics

Component	Stock Solution Concentration	Milliliter Stock Solution/Liter
Macronutrients		
KNO_3	202 g/L	2.5
$Ca(NO_3)_2 \cdot 4H_2O$	236 g/0.5 L	2.5
Iron (sprint 138 Fe chelate)	15 g/L	1.5
$MgSO_4 \cdot 7H_2O$	493 g/L	1
Micronutrients		
H_3BO_4	2.86 g/L	1
$MnCl_2 \cdot 4H_2O$	1.81 g/L	1
$ZnSO_4 \cdot 7H_2O$	0.22 g/L	1
$CuSO_4 \cdot 5H_2O$	0.08 g/L	1
$H_2MoO_4 \cdot H_2O$ or	0.09 g/L	1
$Na_2MoO_4 \cdot 2H_2O$	0.12 g/L	1
Phosphate		
KH_2PO_4 (pH to 6.0)	136 g/L	1

- Make up stock solutions and store in separate bottles.
- Add each component to 800 mL deionized water, then fill to 1 L.
- After the solution is mixed, it is ready to use.

It was F. W. Went in 1957 who first coined the air-growing process as "aeroponics," growing coffee plants and tomatoes with air-suspended roots and applying a nutrient mist to the root section (GreenandVibrant.com, 2019).

The current history of aeroponics is captured in the next chapters that include the Science, Innovation, Business, Practice, and Research. These chapters describe the scope of experiments and findings published in the scientific literature, the intellectual property, and the business literature in the last 70 years.

REFERENCES

Aeroponics: Wikipedia, 2019, https://en.wikipedia.org/wiki/Aeroponics.

Carter, W. A., 1942, A method of growing plants in water vapor to facilitate examination of roots. *Phytopathology* 732: 623–625.

GreenandVibrant.com, 2019, The History of Hydroponics – The Past, The Present, and The Future, www.greenandvibrant.com/history-of-hydroponics.

Grene, M., Depew, D., 2004, *The Philosophy of Biology: An Episodic History*. Cambridge: Cambridge University Press, 11.

Healy, J. F., 2004, *Pliny the Elder: Natural History: A Selection*. London: Penguin Classics.

Hoagland, D. R., Arnon, D. I., 1938, *The Water-Culture Method for Growing Plants without Soil* (Circular (California Agricultural Experiment Station), 347 ed. Berkeley: University of California, College of Agriculture, Agricultural Experiment Station. Retrieved 1 October 2014. www.hdl.handle.net/2027/uc2.ark:/13960/t51g1sb8j.

Hoagland Solution: Wikipedia, 2014, The Hoaglands solution for hydroponic cultivation. *Science in Hydroponics*. Retrieved 1 October 2014. www.en.wikipedia.org/wiki/Hoagland_solution.

Janick, J., Paris, H. S., Parrish, D. C., 2007, The cucurbits of Mediterranean antiquity: Identification of taxa from ancient images and descriptions. *Annals of Botany* 100 (7): 1441–1457.

Rohrbaugh, P. A., 2015, Introduction to Hydroponics and Controlled Environment Agriculture, University of Arizona, https://ceac.arizona.edu/resources/intro-hydroponics-cea.

Royal Netherlands Academy of Arts and Sciences. 2020, Retrieved 20 July 2015, "J. von Sachs (1832–1897)". www.wikidata.org/wiki/Q61650.

Steiner, A. A., 1985, The History of Mineral Plant Nutrition till about 1860 as a source of soil-less culture methods. *Soil-less Culture* 1 (1): 7–24.

Stoner, R. J., 1983, Aeroponics versus bed and hydroponic propogation. *Florist's Review* 1 (173): 4477.

University of Chicago, ed., 1993, "Hanging Gardens of Babylon". Britannica. 5 (15 ed.). Chicago, IL: Encyclopædia Britannica Inc., 681–682.

von Sachs, Julius: Wikipedia, 2019, https://en.wikipedia.org/wiki/Julius_von_Sachs

Zobel R. W., Del Tredici, P., Torrey, J. G., 1976, Method for growing plants aeroponically. *Plant Physiology* 57: 344–346.

3 The Aeroponic Value Proposition

There is an appointed time for everything...a time to plant and time to uproot what is planted.

Eccleisiastes 3:1–2

Value proposition is defined as an innovation intended to make a product attractive to customers. So what makes aeroponics attractive to customers. It is unique. It is creative. It is simple. It produces healthy food. It is universal. It is educational. It can be practiced anywhere in the world. It is easy to learn how to do. It is very productive. It is new. It is fun. We have been talking about many examples of aeroponic systems in the world (Chapter 1) and about its history (Chapter 2). But let's define what it really is.

Aeroponics is best defined by comparison with other agricultural techniques. The most common technique in agriculture is geoponics. Geoponics is the conventional way of growing crops in soil, in the open air, with irrigation, the application of fertilizer (nutrients), pest and weed control. There are three agricultural techniques that are considered to be soil-less approaches. These include hydroponics, aquaponics, and aeroponics. In some cases hydroponics and aquaponics are practiced using media (vermiculite, perlite, pea gravel, sand, expanded clay, pumice, scoria, and polyurethane) that is a substitute for soil but they can also be practiced without media. The roots of the plants are always submerged in nutrient solution. Aeroponics uses a substrate, typically a cube of rockwool (silica dioxide absorbent fibers), for planting seeds and providing a support for the roots and the stem to be anchored in. The roots hang in air.

Conventional agricultural systems use large quantities of irrigation fresh water and fertilizers, with relatively marginal returns (Robinet, 2014). Hydroponics, aeroponics, and aquaponics are modern agriculture systems that utilize nutrient-rich water rather than soil for plant nourishment (Calderone, 2018). Because it does not require fertile land in order to be effective, these new modern agriculture systems require less water and space compared with the conventional agricultural systems. One more advantage of these technologies is the ability to practice vertical farming production which increases the yield per unit area (Technology Spotlight, 2018). The benefits of the new modern agriculture systems are numerous. In addition to higher yields and water efficiency, when practiced in a controlled environment, these new modern systems can be designed to support continuous production throughout the year (Pfeiffer, 2004).

The most clear distinction between aeroponics and hydroponics is how the roots are exposed to the nutrient solution. In hydroponics the roots are submerged in the nutrient solution, whereas in aeroponics the roots are suspended in air and are periodically misted or irrigated with the nutrient solution. Hydroponics can be conducted in many ways—ebb and flow, drip/pass, standing aerated nutrient solution, and nutrient flow technique (Pfeiffer, 2003). The initial development of aeroponics consisted of misting roots using spray nozzles to coat the roots with the nutrient solution. It has also been practiced using vertical columns and pumping the nutrient solution to the top of the column and having it cascade down inside the enclosed column coating the roots with the nutrient solution. Columns have been designed that have both a circular and square profile.

Aquaponics can be conducted in tandem with a hydroponic or aeroponic growing system. Aquaponics uses a tank of live fish to produce a waste solution that is the nutrient solution which is circulated to feed the plants in a hydroponic or aeroponic system. The advantage of aquaponics is mainly from a business standpoint in that the fish can also be sold as well as the produce.

In all cases, these three approaches can be conducted in vertical farming systems. The hydroponic approach requires stacking of horizontal trays on top of one another and requires artificial lighting between trays to replace sunlight. Aeroponic growing can be conducted using vertically oriented columns (towers) either in greenhouses with sunlight or indoors using artificial lights. Aeroponics can also be conducted on horizontal tables with roots misted but typically this does not lend itself to vertical farming.

These approaches compare to geoponics (soil-grown) plants in that the roots are maintained under a dark environment like the soil and the stems and leaves of the plants are exposed to sunlight or artificial light to promote photosynthesis. But essentially for plants to grow they need water-soluble nutrients that can be absorbed by the roots as well as contact with air for the leaves to absorb carbon dioxide which is used in the photosynthesis process of making sugars, starches, and cellulose to ensure plant health and growth. In traditional farming these nutrients are mainly nitrogen bearing ions like nitrate and ammonium, and also phosphorus in phosphate and potassium ions. In fertilizer nomenclature, this is described as NPK for these three nutrients. For soil, these nutrients (fertilizers) are added directly to the soil and when the soil is irrigated or receives rain these ions dissolve in the water and migrate to the root hairs and can be absorbed into the roots for plant growth. The challenge for soil-grown plants is that 60% of the fertilizer distributed on the soil is never absorbed by the plants and ends up becoming runoff, which can become an environmental issue.

In the aeroponic technique, the nutrients (NPK) are added to the water in very low concentrations—parts per million—and can be circulated multiple times past the roots so that they can be absorbed. This is a savings in the use of nutrients because there is a minimal loss in these systems. Also, the roots absorb oxygen and in soil-grown plants the oxygen maybe restricted in diffusing to the roots. In addition to this more efficient use of nutrients without runoff, the quantity of water necessary to grow aeroponically has been estimated to be substantially less than that for soil growing (Pfeiffer, 2003).

Comparing conventional agriculture to aeroponics, it should be noted that one can identify several negative impacts. These would include high and inefficient use of water, large land requirements, high concentrations of nutrient consumption (with lower actual plant consumption), limited crop cycle times dependent on sunlight, weather conditions, and soil degradation. That coupled with the world population growing exponentially seems to suggest that there needs to be a higher rate of production of food.

Therefore, the need for large amounts of high-quality vegetable products to meet the growing demand of this world population seems to justify the development of technologies which optimize the water and nutrient solution demand. The knowledge of water and nutrient uptake by plants is critical for developing control strategies, which increase the possibility to supply the required amounts of water and nutrients for maximum crop growth.

In an aeroponic system, plants can grow without soil and use less water than conventional agriculture. So in aeroponics the water is recycled and is only lost through evaporation or transpiration by the plants. The same is true of the nutrient usage where at least efficiencies of 50% usage compared to conventional techniques with the added benefit of reduction in fertilizer runoff into streams and lakes. The cycle times for most crops can be reduced substantially which again increases productivity and the minimum use of input resources. There are many advantages in growing crops aeroponically:

1. Soil is not necessary
2. Land use is optimized
3. Yields are stable and high with reduced cycle times
4. No nutrient pollution is released into the environment
5. Higher nutrient and water use efficiency due to control over nutrient levels

Aeroponics and hydroponics are similar in the use of the nutrient-rich water, but they are distinctly different. Hydroponics uses certain media other than soil that retains and distributes nutrient-rich water to feed the plants, whereas aeroponics can use either a misting system or a vertical stream of nutrient solution to deliver nutrients. Aeroponics is more suited for vertical growing configurations and uses space more efficiently.

Both hydroponic and aeroponics system allow for flexibility and control of the quality, health and quantity of the vegetable plants and other produce. Whether a hydroponics or aeroponics system has been chosen, both promote self-sustainability in an environmentally friendly way. Hydroponics uses only 10% of water resources when compared to conventional methods giving the grower complete control over nutrient delivery. With aeroponics, there is virtually no grow medium used and a nutrient-rich solution is sprayed on or poured over the root system providing for maximum nutrient absorption. Where an aeroponics system will require constant attention, the hydroponics system may be easier for beginners. However, both systems are much more efficient than soil-based agriculture, and both of them have almost the same opportunity for the flexibility to control the irrigation and nutrient applications.

Aeroponics systems can reduce water usage by 98%, fertilizer usage by 60%, and pesticide usage by 100%, all the while maximizing crop yields. Plants grown in the aeroponics systems have also been shown to uptake more minerals and vitamins, making the plants healthier and potentially more nutritious. Also, since the roots are exposed to the oxygen in the air, the uptake of oxygen is substantially higher.

This aeration not only increases plant growth but also aids in preventing pathogen formation. Aeroponics has the advantage over hydroponics that uses media when transplanting since the plants don't suffer from transplant shock.

One possible challenge with hydroponic and aeroponic growing system is the possibility of water-borne disease traveling rapidly between plants. This is rare and with early detection can be addressed.

In a recent article, the three systems—hydroponics, aeroponics, and aquaponics were compared (Ali AlShrouf, 2017). The author indicated that they have some common similarities since they share the elimination of the soil as a medium to grow crops; with the aim being to deliver sustainable and profitable food production. However, they point out that there are many significant differences.

Comparing aeroponics versus hydroponics, they are both equally efficient at producing healthy and fresh produce. Although aeroponics and hydroponics are similar in the usage of the nutrient-rich water, they are distinctly different. Hydroponics uses certain media other than soil that retains and distributes nutrient-rich water to feed the plants, whereas aeroponics normally uses a misting system to deliver nutrients. Aeroponics succeeds more in vertical growing arrangements and using the space efficiently.

Both hydroponic and aeroponics systems allow for flexibility and control the quality, health, and quantity of the vegetable plants and other produce. Whether a hydroponics or aeroponics system has been chosen, both promote self-sustainability in an environmentally friendly way. Hydroponics uses only 10% of water resources when compared to conventional methods giving the grower complete control over nutrient delivery. With aeroponics, there is virtually no grow medium used and a nutrient-rich solution is sprayed onto the root system providing for maximum nutrient absorption. However, both systems are much more efficient than soil-based agriculture, and both of them have almost the same opportunity for the flexibility to control the irrigation and nutrient applications.

Aeroponics systems can reduce water usage by 98%, fertilizer usage by 60%, and pesticide usage by 100%, all while maximizing crop yields. Plants grown in the aeroponics systems have also been shown to uptake more minerals and vitamins, making the plants healthier and potentially more nutritious. Another benefit is that the plants can more easily be transplanted, since they don't suffer from transplant shock.

Aeroponics also allows one to observe the plants directly without disturbing them, which allows one to adjust the nutrient mix that you're using and cut off any problems that one might be having before they actually have a chance to become a problem. Under the hydroponic system and because the nutrient solution is passed between plants, it is possible for water-borne disease to travel rapidly between them. Also, hydroponic systems, including aeroponics, rely on electricity and require costly generator back-ups to cover for power outages. Hydroponic systems can also be expensive to set up due to the nature of the equipment involved. However, once the system is, it is cheaper than a conventional farming to operate. Since the plant roots are isolated and there is no planting medium used in the aeroponics system, plants that are grown with this suspended, misted system will get maximum nutrient absorption.

Aeroponics systems are favored over other methods of hydroponics because the increased aeration of nutrient solution delivers more oxygen to plant roots, stimulating growth and preventing pathogen formation.

The deciding factors in choosing one method over the other are the ancillary benefits. For aeroponics, this includes lower water and nutrient usage. For certain hydroponic systems, this includes greater buffering capacity and room for error.

In summary, the main advantage of these modern cultivation systems is the conservation of water which increases productivity per unit area. While all three, hydroponic, aeroponic, and aquaponic, can be implemented in a raised garden, all three are very similar in every way except hydroponics and aeroponics requires the addition of fertilizer and there's no fish in the nutrient solution. In aquaponics, plants and fish live a symbiotic life with the fish feeding the plants, and the plants cleaning and filtering the fish's environment. With the advent of process control sensors that automatically monitor and adjust pH and nutrient levels, the maintenance requirements for aeroponics has become much simpler.

Here is a summary of the upsides and downsides to aeroponic growing.

UPSIDES

No soil is needed. One can grow crops in places where the land is limited, doesn't exist, or is contaminated. Hydroponics have been around for centuries. Most recently fresh vegetables were grown in the 1940s on Wake Island for the troops in the Pacific Ocean. In the 1990s, it was applied to grow food in the MIR space station to feed the astronauts.

It makes a better use of space and location. It is flexible. One can grow plants in a small apartment in a spare area. In soil-grown plants, the roots spread out to find water and nutrients including oxygen. In these systems, the roots are in contact with exactly the nutrients that they need and all the oxygen they can absorb. This translates into plants growing much closer to each other, the reduction in wasted space, and faster cycle times.

Labor for tilling, cultivating, fumigating, watering, and other traditional practices is largely eliminated.

The climate can be controlled either in a greenhouse or even in a storage container. The key variables can be monitored and automatically tuned. This includes temperatures, humidity, light intensity, carbon dioxide concentration, nutrient concentrations, and air composition. This translates into being able to grow foods all year round regardless of the season. Farmers can control when they want to produce certain crops based on anticipated demands.

Aeroponics saves water. Aeroponically growing systems use less than 10% of the water that is necessary for soil-based growing. Plants as they grow will take up more and more water as necessary, where the water used in soil-based systems tends to migrate and large losses occur. Aeroponic systems lose water through evaporation, transpiration, and leaks but present systems have reduced that to a minimum.

In nutrient usage, the aeroponic growing system can be designed to control 100% of the nutrients that the plants need. Since the mechanism for feeding the plants is by transfer of the nutrient ions (nitrates, phosphates, potassium, and other ions) by the root hairs to the stem of the plant, this approach allows the plants to take up all the nutrients they need, when they need them. Nutrients are also conserved in the feed tanks so that there are no losses or changes of nutrients like there may be in the soil. It is a reproducible system that is not dependent on weather, rain, and outdoor temperatures.

The pH control of the nutrient solution or the effective hydrogen ion concentration can be controlled and monitored to ensure proper concentration of nutrients are soluble in the water and available to the roots. The pH is easily controlled as compared to soils.

The growth rate of plants grown aeroponically is faster because all the inputs—temperature, light, moisture, and nutrients are controlled accurately. Plants don't need to waste valuable energy searching of diluted nutrients in the soil. They shift all of their focus on growing and producing fruit.

There are no weeds to contend with as is the case with soil-grown plants. It also saves time that is consumed controlling weeds in soil. Weeds also compete with the plants for the nutrients. That does not happen in these systems.

Eliminating soil helps make plants less vulnerable to soil-borne pests like birds, gophers, groundhog: and diseases like Fusarium, Pythium, and Rhizoctonia species. Using the latest cloud systems, one can easily take control of most key variables.

Again eliminating soil means using less insecticides and herbicides this reduces plant diseases and there are fewer chemicals. This translates into growing cleaner and healthier foods. This is one of the biggest health benefits that address the concern for food safety. Soil-borne plant diseases are more readily eradicated in closed systems.

This growing system reduces labor time and saves time since the need for tilling, watering, cultivating, applying herbicides and pesticides is reduced or eliminated.

More complete control of the environment, timely nutrient feeding or irrigation and in a greenhouse-type operations, the light, temperature, humidity, and composition of the air can be manipulated.

Aeroponics reconnects people with food. This is a lost experience due to our current food production and distribution systems. The amateur horticulturist can even adapt an aeroponics system to home and patio-type gardens even in high-rise buildings. An aeroponic system can be clean, lightweight, and mechanized.

DOWNSIDES

Aeroponic growing requires time and attention to detail. Focus, diligence, and effort pays off in satisfying yields and nutritious food. However in soil-based growing, plants can be left on their own for days, and they will still survive for short periods. The soil chemistry and weather conditions help to regulate and maintain balance. That is not the case for aeroponics. Plants will die out more quickly without proper care and adequate knowledge. Automation can be implemented in these systems to ensure proper operation.

Experience and technical knowledge is necessary which requires specific expertise for devices used, plants grown, what they need to survive. Mistakes could be made that could cause deleterious effects to your growing system. Trained personnel must direct the growing operation. Knowledge of how plants grow and of the principles of nutrition are important.

There are some concerns about whether aeroponics should be certified as organic growing. Some question whether plants grown this way will get microbiomes that are in the soil. But people in the US, Australia, Japan, and Holland have provided food for millions of people. There are trade-offs. With soil the risks are related to pesticide and herbicide residues and pests. It has been demonstrated that some aeroponic systems can use microbiomes. Current research is underway to address this concern.

These systems require a water source and electricity. The water source can be rainwater that can be collected and saved prior to use. Electricity is required but could be supplied by solar panels in areas where electricity is not easily accessible.

System failures can be a threat. These failures can be caused by electric outages which affect the pumps from circulating the nutrient solutions to the roots. If this happens plants may dry out quickly and will die in several hours. Backup power may be a consideration to address unexpected outages for large commercial operations.

Initial expenses are relatively high compared to soil growing and are dependent on the size of the garden. Systems require containers, lights, a pump, a timer, and nutrients. Once the system is functioning, the cost will be reduced to only nutrients and electricity.

There is a long return on investment depending again on the scale of the system. This is largely because of the high initial expenses and the long, uncertain ROI, return on investment.

Growing plants in a closed system using water means that plant infections can escalate fast to plants on the same nutrient reservoir. In most cases, diseases and pests are not so much of a problem in a small system of home growers. Should the diseases happen, one should sterilize the infected water, nutrient, and the whole system as fast as possible (Jones, 2005).

REFERENCES

AlShrouf, A., 2017, Hydroponics, Aeroponics, and Aquaponics as Compared with Conventional Farming. *American Scientific Research Journal for Engineering, Technology, and Sciences (ASRJETS)* 27 (1): 247–255.

Calderone, L., 2018, Indoor and Vertical Farming, Monitoring, and Growing. agritechtomorrow.com, 12/26/2018.

Jones Jr., J. B. (ed.), 2005, *Hydroponics: A Practical Guide for the Soil-less Grower.* Boca Raton, FL: CRC Press, 4–5.

Pfeiffer, D. A., 2003, *Organic Consumers Association: Eating Fossil Fuels.* Silver Bay City, MI: The Wilderness Publisher.

Pfeiffer, D. A., 2004, *Eating Fossil Fuels.* Gabriola: New Society Publishers. www.organicconsumers.org/news/eating-fossil-fuels-dale-allen-pfeiffer.

Robinet, R., 2014, Sustainable versus Conventional Agriculture. https://you.stonybrook.edu/environment/sustainable-vs-conventional-agriculture/.

Technology Spotlight, Sustainability in Three Dimensions, 2018, https://modernag.org/innovation/benefits-vertical-farming-robotics/.

4 Aeroponic Science

The most beautiful thing we can experience is the mysterious. It is the source of all true art and science.

Albert Einstein

In the book, *Hydroponics*, published in 2005, the author writes about the future of hydroponics (Jones, 2005). What he says is most likely true of aeroponics as well. He writes "What is not encouraging for the future is the lack of input from scientists in public agricultural colleges and experimental stations that at one time made significant contributions to crop production procedures, including hydroponics. He talks about the earlier researchers, Gericke and Hoagland at the University of California. He indicates at the time of the publishing of his book that only a few researchers at the university were still active in hydroponic investigations and research. Fortunately, since the early 2000s the pace of aeroponic research has begun to increase. Also, the number of both US and international university research programs is expanding. This is partially true because of the need for better methods of food production but also because aeroponic systems lend themselves to conducting definitive research. This is due to the fact that essentially these systems are closed systems. One can monitor the uptake of nutrients by measuring the concentrations of them in the nutrient solution and at the same time measuring tissue samples to determine the levels of these nutrients that end up in the tissue. One can also control the pH, water and air temperature, the carbon dioxide concentration as well as nutrient concentrations. The author also expresses concerns about the decline in the hydroponic organizations like the Hydroponic Society of America. He hoped that the internet will be useful in increasing research in this area.

The expansion of the field of aeroponics is a relatively recent development. Hydroponics is much more developed and has been in existence for many more years. One measure of the "coming of age" of a technology is to plot the number of academic journal articles (peer-reviewed technical literature) published by year for the last 50 years. After conducting a literature search, Figure 4.1 was compiled to plot this trend. Since 1975, the trend has been steadily increasing and in the last 5–10 years it has increased 3-fold. Technical papers were identified having the word "aeroponic" in its title or keyword section. In reviewing topics of interest to the research community of these 238 papers, the top eight topic in terms of the number of papers published were found to be:

potatoes	35
technology	18
roots	14
nutrients	13
medicinal	12
maize	12
lettuce	12
fungi	12

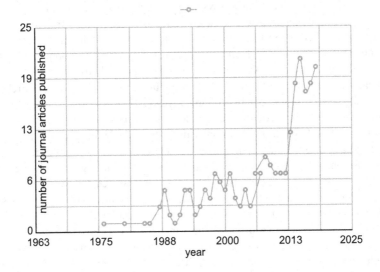

FIGURE 4.1   The trend of the number of technical papers published about aeroponics in the last 50 years.

Other papers covered several other crops and topics:

Acacia
Alfalfa
alpine penny cress
antioxidants
arugula
asparagus
barley
basil
begonia
biomass
blackberry
broccoli
camphor

carbon dioxide
carpetweed
chickpea
Chinese cabbage
Chrysanthemum
Corn
Cowpea
Cranberry
Cucumber
Elm
Eucalyptus
Evergreen
Fir
food security
grape
iris
lotus
maize
muskmelon
olive
pea
peanut
pepper
petunia
radish
review
rice
saffron
seeds
shallot
social impact
soybean
space applications
spruce
strawberry
sunflowers
tomato
trees
vegetables
wheat
yams

These papers are listed with a brief abstract describing the research. This summary demonstrates a very broad scope of potential applications for aeroponic growing. It also demonstrates the global interest in this technology in that all continents have countries investing in research in this food technology. These papers have been organized by topic, arranged alphabetically, and include the country where the research was conducted, the title, principal author, and the abstract. The references for these papers are provided in the references at the end of the chapter.

ACACIA

Singapore/France
Aeroponic Production of *Acacia mangium* Saplings Inoculated with arbuscular mycorrhiza (AM) Fungi for Reforestation in the Tropics (Martin-Laurent et al., 1999).

> *Martin-Laurent, Fabrice et al., National Institute of Education, Nanyang Technological University, Singapore and the Centre de Coopération Internationale en Recherche Agronomique pour le Développement, Cedex, France.*

This paper describes an aeroponic, a soil-less plant culture method for the production of *Acacia mangium* saplings associated with AM fungi. *A. mangium* seedlings were first grown in multipots and inoculated with Endorize, commercial AM fungal inoculum. They were then, either transferred to aeroponic systems or to soil. Aeroponics was found to be a better system than soil, allowing the production of tree saplings twice as high as those grown in soil. Moreover, compared to plants grown in soil, aeroponically grown saplings inoculated with AM fungal inoculum exhibited significantly different rates of mycorrhization, resulting in an increase in phosphorus and chlorophyll in plant tissues. Their results suggest that the aeroponic system is an innovative and appropriate technology which has the potential to produce in large quantities, tree saplings associated with soil micro-organisms, such as AM fungi, for reforestation of the degraded land in the humid tropics.

Singapore/France
A New Approach to Enhance Growth and Nodulation of *Acacia mangium* through Aeroponic Culture (Martin-Laurent et al., 1997).

> *Martin-Laurent, Fabrice et al., National Institute of Education, Nanyang Technological University, Singapore and the Centre de Coopération Internationale en Recherche Agronomique pour le Développement, Cedex, France.*

This study was conducted using aeroponics as an alternative method to classical soil inoculation procedures for the production of hypernodulated legume tree saplings. The study was designed to determine whether a plant culture method on non-solid media could be used as an alternative for inoculation of *Acacia mangium* with selected strains of Bradyrhizobium spp. *A. mangium* seedlings were grown and inoculated with Bradyrhizobium strain Aust13c and strain Tel2 in hydroponics, aeroponics, and sand. Aeroponics was found to be the best system of the three, allowing the production of tree saplings 1 m in height after only 4 months in culture. Moreover, compared to plants grown in liquid or sand media, aeroponically grown saplings inoculated with Bradyrhizobium spp. developed a very high number of small nodules distributed all along the root system, resulting in an increase in the nitrogen and chlorophyll content in plant tissues.

Malaysia

Effects of Nitrogen Source on the Growth and Nodulation of *Acacia mangium* in Aeroponic Culture (Weber et al., 2007).

Weber, J. et al., Universiti Putra Malaysia Putrajaya, Selangor Darul Ehsan, Malaysia.

This paper describes a study of the effects of ammonium and nitrate on growth and nodulation rates of *Acacia mangium* inoculated with Bradyrhizobium and grown in aeroponic culture. Concentrations of 13.6 and 4.9 mM, nitrate stimulated plant growth, nitrogen uptake, and the total chlorophyll content compared with corresponding concentrations of ammonium, which had a deleterious effect. On the other hand, nodulation was depressed with nitrate and totally suppressed with ammonium at these two concentrations. However at 0.4 mM, ammonium actually stimulated nodulation rates and resulted in robust plant growth comparable to that obtained with higher nitrate concentrations. Ammonium nitrification was confirmed to be absent from measurements of the nutrient solutions in the aeroponic culture tanks.

France

Survival and Growth of *Acacia mangium* Wild Bare-Root Seedlings after Storage and Transfer from Aeroponic Culture to the Field (Weber et al., 2005).

Jean J W Weber et al., Nanyang Technology University/National Institute of Education, Natural Science Division, Singapore; Laboratoire des Symbioses Tropicales et Méditerranéennes, Cedex, France. UMR INRA-University Henri Poincaré Nancy, Interactions Arbres/ Microorganismes, Faculté des Sciences, Cedex, France.

This paper demonstrated experimentally that aeroponic culture was a promising nursery technique to raise *A. mangium* and to improve growth rates as well as to control the level of infection with rhizobia and mycorrhizal fungi. This work was designed to determine whether aeroponically grown bare-root seedlings can be stored out of aeroponic troughs, and/or planted to the field without acclimatization in polybags. After field planting, no significant differences in terms of survival and growth rates were expressed between bare-root seedlings that had been stored in plastic bags for 6 days or directly transferred to the field, or acclimatized in polybags. Aeroponic culture appears to be the method of choice to obtain high-quality seedlings, which are much easier to plant and transport compared to those obtained under classical nursery techniques using soil or solid substrate.

ALFALFA

Morocco

Variations in leaf gas exchange, chlorophyll fluorescence, and membrane potential of *Medicago sativa* root cortex cells exposed to increased salinity: The role of the antioxidant potential in salt tolerance (Farissi et al., 2018).

Farissi, Mohamed et al., Faculté des Sciences et Techniques Guéliz Equipe de Biotechnologie Végétale et Agrophysiologie des Symbioses Marrakech, Morocco.

This paper describes a study of the effects of salinity on some ecophysiological and biochemical criteria associated with salt tolerance in two Moroccan alfalfa (*Medicago sativa* L.) populations, Taf 1 and Tata. Salinity is one of the most serious agricultural problems that adversely affects growth and productivity of pasture crops such as alfalfa. The experiment was conducted in a hydro-aeroponic system containing nutrient solutions, with the addition of NaCl at concentrations of 100 and 200 mM. The salt stress was applied for a month. Several traits in relation to salt tolerance, such as plant dry biomass, relative water content, leaf gas exchange, chlorophyll fluorescence, nutrient uptake, lipid peroxidation, and antioxidant enzymes, were analyzed at the end of the experiment. The Tata population was more tolerant to high salinity (200 mM NaCl) and its tolerance was associated with the ability of plants to maintain adequate levels of the studied parameters and their ability to overcome oxidative stress by the induction of antioxidant enzymes, such as guaiacol peroxidase, catalase, and superoxide dismutase.

USA

Elevated Carbon Dioxide Concentration around Alfalfa Nodules Increases Nitrogen Fixation (Fischinger et al., 2010).

Stephanie A. Fischinger et al., Ragon Institute of MGH, MIT, and Harvard, Cambridge, MA, USA.

This paper describes a study of nitrogen fixation in alfalfa plants when the nodules are exposed to elevated carbon dioxide concentrations. Nodule carbon dioxide fixation is known to depend on external carbon dioxide concentration. Therefore, nodulated plants of alfalfa were grown in a hydroponic and aeroponic systems that allowed separate aeration of the root/nodule compartment and avoided any gas leakage to the shoots. More intensive carbon dioxide and nitrogen fixation coincided with higher per plant amounts of amino acids and organic acids in the nodules. Moreover, the concentration of asparagine was increased in both the nodules and the xylem sap. The data support the thesis that nodule carbon dioxide fixation is pivotal for efficient nitrogen fixation. It was concluded that sufficient carbon dioxide application to roots and nodules is necessary for growth and efficient nitrogen fixation in hydroponic and aeroponic growth systems.

ALPINE PENNY-CRESS

France

Cadmium Uptake and Partitioning in the Hyperaccumulator *Noccaea caerulescens* Exposed to Constant Cd Concentrations throughout Complete Growth Cycles (Lovy et al., 2013).

Lovy, Lucie et al., Université de Lorraine, INRA, Laboratoire
Sols et Environnement, Cedex, France.

The cadmium (Cd) hyperaccumulation kinetics were studied in the different plant organs, throughout the complete cultivation cycle, independently of a possible soil effect. Plants of *Noccaea caerulescens* were exposed in aeroponics to three constantly low Cd concentrations and harvested at siliquae formation. Dry matter allocation between roots and shoots was constant over time and exposure concentrations, as well as Cd allocation. However, 86% of the Cd taken up was allocated to the shoots. Senescent rosette leaves showed similar Cd concentrations to the living ones, suggesting no redistribution from old to young organs. The Cd root influx was proportional to the exposure concentration and constant over time, indicating that plant development had no effect on this. The bio-concentration factor (BCF), i.e., [Cd]/[Cd] for the whole plant, roots or shoots was independent of the exposure concentration and of the plant stage. Cadmium uptake in a given plant part could therefore be predicted at any plant stage by multiplying the plant part dry matter by the corresponding BCF and the Cd concentration in the exposure solution.

AONLA

India

Effect of Micronutrients on Growth and Yield of Aonla (*Emblica officinalis* gaertn.) cv. NA-7 (Abhijith et al., 2017).

Abhijith, Y.C. et al., Dept. of Fruit Science, College
of Horticulture, Bengaluru, India.

This study was conducted to assess the response of foliar application of different micronutrients on growth and yield of aonla cv. NA-7. Aonla (*Emblica officinalis* Gaertn.) is one of the most important minor fruits of India, which is also known as Indian gooseberry which belongs to the family Euphorbiaceae and is native to Central and Southern India. Though it is a hardy crop, growers are experiencing the problem of heavy premature fruit drop leading to reduced yield and sometimes reduced quality due to necrosis which may be due to deficiency of nutrients, particularly micronutrient. The results revealed that the foliar spray of micronutrients combination of 0.5% zinc sulfate + 0.5% iron sulfate + 0.25% borax significantly increased overall growth of plants, reduced the incidence of fruit drop (45.60% as against 79.63% in control) resulting in increased fruit set (53.73% as against 21%

in control). The said combination of micronutrients was also associated with highest fruit weight (43.69 g), fruit length (3.78 cm), fruit diameter (4.93 cm), and yield (24.96 kg/plant).

ANTIOXIDANTS

USA
Assessment of Total Phenolic and Flavonoid Content, Antioxidant Properties, and Yield of Aeroponically and Conventionally Grown Leafy Vegetables and Fruit Crops: A Comparative Study (Chandra et al., 2014).

> *Chandra, Suman et al., National Center for Natural Product Research, Research Institute of Pharmaceutical Sciences, School of Pharmacy, University of Mississippi, Oxford, USA.*

This study was conducted to compare product yield, total phenolics, total flavonoids, and antioxidant properties in different leafy vegetables/herbs (basil, chard, parsley, and red kale) and fruit crops (bell pepper, cherry tomatoes, cucumber, and squash) grown in aeroponic growing systems (AG) and in the field (FG). An average increase of about 19%, 8%, 65%, 21%, 53%, 35%, 7%, and 50% in the yield was recorded for basil, chard, red kale, parsley, bell pepper, cherry tomatoes, cucumber, and squash, respectively, when grown in aeroponic systems, compared to that grown in the soil. Antioxidant properties of AG and FG crops were evaluated using 2,2-diphenyl-1-picrylhydrazyl (DDPH) and cellular antioxidant (CAA) assays. In general, the study shows that the plants grown in the aeroponic system had a higher yield and comparable phenolics, flavonoids, and antioxidant properties as compared to those grown in the soil.

ARUGULA

Mexico
PRODUCCIÓN ACUAPÓNICA DE TRES HORTALIZAS EN SISTEMAS ASOCIADOS AL CULTIVO SEMI-INTENSIVO DE TILAPIA GRIS (*Oreochromis niloticus*) (Ronzón-Ortega et al., 2015).

> *Ronzón-Ortega, M. et al., Instituto Tecnológico de Boca del Río, División de Estudios de Posgrado e Investigación. Laboratorio de Mejoramiento Genético y producción Acuícola, Veracruz, México.*

Three production systems were studied for edible plants, arugula (*Eruca vesicaria*), cilantro (*Coriandrum sativum*), and tomato (*Solanum lycopersicum*), associated with the semi-intensive cultivation of tilapia (*Oreochromis niloticus*), in order to determine their adaptation and productive efficiency. A completely random experimental design was used, where three techniques for aquaponics were tested for plant production: Aqua-aeroponics system (SAC1); Aquaponics system with a porous and inert substrate (SAC2); Aquaponics system with solid rain as the fixating substrate

(SAC3); the following were cultivated simultaneously: arugula, tomato, and cilantro. The growth results for the three plant varieties, stem length, number of leaves and ramifications, both in SAC2 and SAC3, were efficient, particularly in SAC2 where the arugula and tomato plants with highest growth were found, although not significantly different between treatments; the cilantro plants cultivated in SAC3 had the highest growth. In contrast, the three varieties of plants cultivated in SAC1 presented lower survival and growth.

Colombia

Automatic Aeroponic Irrigation System based on Arduino's Platform (Montoya et al., 2017).

A P Montoya et al., Universidad Nacional de Colombia, Medelln, Colombia.

This paper describes the development of an automatic monitored aeroponic-irrigation system based on the Arduino's free software platform. The recirculating hydroponic culture techniques, as aeroponics, has several advantages over traditional agriculture, and is aimed to improve the efficiently and environmental impact of agriculture. These techniques require continuous monitoring and automation for proper operation. Analog and digital sensors for measuring the temperature, flow and level of a nutrient solution in a real greenhouse were implemented. In addition, the pH and electric conductivity of nutritive solutions are monitored using the Arduino's differential configuration. The sensor network, the acquisition and automation system are managed by two Arduinos modules in master-slave configuration, which communicate with each other by Wi-Fi. Further, data are stored in micro SD memories and the information is loaded on a web page in real time. The developed device brings important agronomic information when it was tested with an arugula culture (*Eruca sativa* Mill). The system could also be employed as an early warning system to prevent irrigation malfunctions.

ASPARAGUS

Poland

The effect of temperature and crown size on asparagus yielding (Gąsecka et al., 2009).

Gąsecka, Monika et al., of Chemistry, Poznan
University of Life Sciences, Poznan, Poland.

The paper describes a study of the effect of temperature on asparagus yields of different crown sizes, planted in growth chambers in an aeroponic system with recirculation. The results showed that asparagus yield was dependent on air temperature and crown size; however, crown size had a greater influence on the yield. The diameter and weight of the asparagus spears were also dependent on crown size. Higher dry weight content, degrees Brix, fructan, and total carbohydrate content in storage roots were documented in large crown asparagus plants before and after harvest.

BARLEY

Germany/USA

Infection of Barley Roots by *Chaetomium globosum*: Evidence for a Protective Role of the Exodermis (Reissinger et al., 2003).

Reissinger, Annette et al., Soil Ecosystem Phytopathology, Institute for Plant Diseases, University of Bonn, Germany and the Department of Microbiology, University of Pennsylvania School of Medicine, Philadelphia, PA, USA.

This paper involved a study of the infection of barley roots using Murashige and Skoog (MS)-agar and aeroponic culture as axenic plant growth systems. *Chaetomium globosum* pathogenesis was analyzed with serological and histological methods. Irrespective of the growth system, *C. globosum* infected the root epidermis. Roots grown in MS-agar were extensively colonized intercellularly and intracellularly up to the inner cortex and the tissue underwent necrosis. In contrast, roots grown in aeroponic culture were not colonized beyond the epidermis and the roots appeared healthy. The results indicated that specific environmental conditions are important for infection and disease expression in barley roots.

China

Root Border Cell Development is a Temperature-Insensitive and Al-Sensitive Process in Barley (Pan et al., 2004).

Jian-Wei Pan et al., State Key Laboratory of Plant Physiology and Biochemistry, College of Life Sciences, Zhejiang University, Hangzhou, China.

This team conducted in vivo and in vitro experiments that showed that border cell (BC) survival was dependent on root tip mucigel in barley (*Hordeum vulgare* L. cv. Hang 981). In aeroponic culture, BC development was an induced process in barley, whereas in hydroponic culture, it was a kinetic equilibrium process during which 300–400 BCs were released into water daily. The response of root elongation to temperatures (10°C–35°C) was very sensitive but temperature changes had no substantial effect on barley BC development. These results suggested that BC development was a temperature-insensitive but Al-sensitive process, and that BCs and their mucigel played an important role in the protection of root tip and root cap meristems from Al toxicity.

Kazakstan

Technology of mass multiplication of cereal aphids (*Schizaphis graminum*) using an aeroponic plant and dilution of the bioagent aphidius (*Aphidius matricariae*).

Duisembekov, B. et al at the Kazakh Research Institute for Plant Protection.

The results of this research are given on the cultivation of fodder plants of barley and infection of plants with cereal aphids in the conditions of an aeroponics installation. The germination parameters are determined depending on the periodicity of the water supply of its volume and the mass of the seeds grown in the plant. In the conditions of the aeroponic plant, the reproduction of aphids is considered optimal if five individuals of phytophagous are released per barley plant. After 7 days, the number of aphids increased to 42.5 individuals, while its high concentration was noted. When carrying out the infection of aphids propagated under the conditions of the aphids, the optimal parasite ratio: host = 1:60. The degree of infection of aphids (mummified) was 84.2% on this variant.

BASIL

Greece

Nitrogen Nutrition Effect on Aeroponic Basil (*Ocimum basilicum* L.) Catalase and Lipid Peroxidation (Zervoudakis et al., 2015).

Zervoudakis, George et al., Department of Greenhouse Crops and Floriculture; Technological Institute of Mesologgi, Greece.

This study investigated the effect of three different nitrogen nutrition solution concentrations (1.8, 3.6, and 11.5 mM) on leaf and root oxidative stress of aeroponically cultured basil (*Ocimum basilicum* L.) plants. Catalase (CAT) activity and lipid peroxidation (LP) were used as oxidative stress indexes at two different growth stages (10 and 15-week-old plants, respectively). Leaf and root CAT activity was enhanced by the increment of nitrogen concentration at both growth stages of the plants. Especially in younger, high nitrogen nourished plants, 130% and 149% increments of the leaf and root CAT activities were observed, respectively, in comparison with the low nitrogen nourished ones. These results suggest that increased nitrogen nutrition induces oxidative stress mainly in the leaves of aeroponically grown basil plants while the increase in CAT activity probably represents a part of the plant's antioxidative defense against potent cellular damage similar to membrane lipid peroxidation.

Greece

Yield and Nutritional Quality of Aeroponically Cultivated Basil as Affected by the Available Root-Zone Volume (Salachasa et al., 2015).

Salachasa, Georgios et al., Department of Agricultural Technology, Laboratory of Plant Physiology and Nutrition, T.E.I. of Western Greece.

This paper investigated the effect of the available root zone volume on yield and quality characteristics of aeroponically cultivated sweet basil (*Ocimum basilicum*, L.) plants. Growth and photosynthesis were also evaluated. At a fully automated glasshouse aeroponic growing system, plants were cultivated in canals with 10 m length, 0.67 m width for depths: of 0.15 m, 0.30 m, and 0.70 m. Plants cultivated in growing canals with the lower depths 0.15 m and 0.30 m, gave increased dry biomass production; plant height; root length; leaves per plant; total chlorophyll content; net photosynthesis rate; transpiration rate and stomatal conductance, in comparison with plants cultivated in canals with the maximum depth of 0.70 m. In contrast, plants cultivated in 0.70 m depth canals showed statistically increased root dry biomass production. The results showed that basil plants grown aeroponically have superior nutritional quality characteristics.

BEAN

Israel

Allometric Relationships in Young Seedlings of Faba Bean (*Vicia faba* L.) Following Removal of Certain Root Types (Eshel et al., 2001).

A. Eshel et al., Department of Botany, the George S. Wise Faculty of Life Sciences, Tel Aviv University, Israel.

The paper was a study of sink-source relationships and allometric ratios in young seedlings of faba bean (*Vicia faba* L.) following pruning of some root types. The plants were grown in an aeroponic system allowing an easy access to each part of the root system, throughout the experiment, without disturbing the others. Root, leaf, and stem growth as well as their mineral content were determined in one group of undisturbed plants (CTRL) and in four groups of plants treated as follows: TAP—the distal-free portion of the taproot was removed; HALF—half the laterals were removed; ALL—all lateral roots were removed, and TAP+ HALF—both the distal part of the taproot and half of the laterals were removed. The allometric relationships between the surface area of the roots and that of the leaves were restored within the experimental period, apparently due to reduction in shoot growth. Removal of the distal parts of the taproot did not cause an increase in shoot growth. This indicates that the strength of the sinks (mostly of lateral roots) rather than that of the source determines these relationships.

BEGONIA

Sweden

Feature Article: Transpiration Rate in Relation to Root and Leaf Growth in Cuttings of *Begonia* X *hiemalis* Fotsch (Ottosson et al., 1997).

Ottosson, B. et al., Swedish University of Agricultural Sciences, Department of Horticultural Science, Alnarp, Sweden.

The team studied the cuttings of *Begonia* X *hiemalis* Fotsch. cv. 'Schwabenland Red' rooted in an aeroponics system at 21°C and 2.9 mol/(m²day) photosynthetic photon flux density (PPFD) for 18 h/day. Leaf length increase rate was higher in leaves appearing at time of root formation compared with leaves starting to expand before roots were formed. Transpiration rate per unit leaf area increased after roots had been formed. In cuttings potted shortly after root formation, transpiration rate per unit leaf area decreased during the first days after potting and remained at a low level for a week. Leaf area expansion and root growth rate were slowed down during this period. Transpiration per cutting was less correlated with leaf area 2 weeks after root formation in the potted cuttings compared with nonpotted cuttings of the same developmental stage. Cuttings with large leaf area produced greater root mass in relation to cuttings with smaller leaf area.

BIOMASS

USA

Evaluation of Algal Biomass Production on Vertical Aeroponic Substrates (Johnson et al., 2015).

Johnson, Michael et al., Department of Environmental Sciences,
Rutgers University, New Brunswick, NJ, USA.

A novel aeroponic substrate-based cultivation system was studied to determine whether it could produce significant quantities of biomass without a negative impact on the lipid productivity and fatty acid profile compared to the two traditional systems. Large scale algal biomass production have focused primarily on the Open Pond (OP) and Photobioreactor (PBR) systems, but neither system had been able to produce algae biofuel in a financially viable manner. This vertical aeroponic substrate system produced significant areal yields resulting in reduced energy inputs and increased financial return. In addition to productivity increases, the aeroponic nature of this substrate system did not negatively affect the fatty acid composition of the cultivated biomass, thus demonstrating the promising potential for using substrate-based systems to produce biofuel, nutraceuticals, and feed for fisheries and various other applications.

Poland

Concept of Aeroponic Biomass Cultivation and Biological Wastewater Treatment System in Extraterrestrial Human Base (Jurga et al., 2018b).

Jurga Anna et al., Wroclaw University of Science and Technology,
Faculty of Environmental Engineering, Wroclaw, Poland.

This paper describes a study of the concept of an aeroponically based biomass cultivation and a wastewater treatment system in future Planetary Base (PB), e.g., Moon or Mars, designed for an eight-person crew. These two subsystems are part of Life Support System (LSS), which aims at providing proper environmental condition for human habitation.

Iran

Effects of Cultivation Systems on the Growth and Essential Oil Content and Composition of Valerian (Tabatabaei 2008).

Tabatabaei, Seyed Jalal, University of Tabriz,
Department of Horticulture, Tabriz, Iran.

This study assessed the growth and essential oil production of valerian (*Valeriana officinalis* L. var. common) growing in aeroponic, floating, growing media (a perlite and vermiculite mix), and soil systems by measuring biomass production and essential oil content and composition. The highest fresh weight of both leaves (802 g plant[-1]) and roots (364.5 g plant[-1]) was obtained in the floating media system. No significant difference in leaf area between the floating and growing media systems was observed, but comparative leaf area was reduced considerably in the aeroponics and soil systems. Both photosynthesis and stomatal conductance were increased in the floating and growing media systems, as compared with the aeroponics and soil systems, along with the concentration of essential oil. The major constituents of essential oil were bornyl acetate, valerenal, comphene, trans-caryophyllene, cis-ocimen, α-fenchen, and δ-elemene, although the relative proportion of each constituent varied with treatment. The concentration of bornyl acetate was highest (32.1% of total oil) in the floating system, sonic 56.5% higher than the concentration in the soil. The results suggest that under a controlled environment, both floating and growing media systems could be promising approaches for obtaining higher root yields and oil productions in valerian.

BLACKBERRY

Bulgaria

Direct ex Vitro Rooting and Acclimation in Blackberry Cultivar 'Loch Ness' (Fira et al., 2012).

Fira, A. et al., Industrial Plants OOD, Micropropagation, Kazanlak, Bulgaria.

This study was conducted on the direct ex vitro rooting and acclimation experiments in blackberry (*Rubus fruticosus*), thomless cultivar 'Loch Ness'. The plant material consisted of plants propagated on Murashige & Skoog (MS) medium with 0.5 mg/L benzyladenine (BAP). The shoots excised from the plantlets were rooted directly ex vitro in various substrates: floating perlite, plastic sponge inserted in floating cell trays, rockwool in plastic trays covered with transparent lids, as well as potting mixes available commercially: Florasol, Sol Vit G, Florimo. These experimental variants yielded good results regarding the rooting and acclimation percentages. Rooting in Jiffy pellets placed in floating cell trays as well as the use of rockwool in noncovered plastic trays yielded negative results. The experiments regarding ex vitro rooting and acclimation in aeroponics or by suspending the shoots in air saturated with vapor also yielded negative results.

BROCCOLI

Singapore

Interaction Between Iron Stress and Root-Zone Temperature on Physiological Aspects of Aeroponically Grown Chinese Broccoli (He et al., 2008).

He, Jie et al., National Institute of Education, Nanyang Technological University, Singapore.

This team studied the growth of Chinese broccoli (*B. alboglabra*), a subtropical vegetable where the root-zone temperature (RZT) was set at 25°C while its aerial portions were exposed to the hot, fluctuating temperatures for 4 weeks in the tropical greenhouse. Interaction between iron (Fe) stress and RZT were then studied by exposing the plant roots to two different RZTs: a constant cool 25°C-RZT (CRZT) and a fluctuating hot ambient RZT (ARZT). There were three different Fe levels [full Fe (FFe), 1/2Fe, and 0Fe] in the nutrient medium supplied to plants at each RZT. Compared to plant grown at CRZT, hot ARZT resulted in decreases in shoot and root productivities, photosynthetic carbon dioxide assimilation rate (A), stomatal conductance (g s), Fe and nitrate (NO_3^-) uptake and transport and nitrate reductase activity (NRA). Hot ARZT also altered root morphology. These results indicated that hot ARZT may mask the effects of Fe stress on certain physiological process which was clearly elucidated at RZT.

CAMPHOR

Israel

From America to the Holy Land: Disentangling Plant Traits of the Invasive *Heterotheca subaxillaris* (Lam.) (Sternberg, 2016).

Sternberg, Marcelo, Tel Aviv University | TAU School of Plant Sciences & Food Security, Israel.

This paper describes a study of camphor-weed (*H. subaxillaris*) from native (US) versus introduced (Israel) populations to identify functional traits that accorded this species invasion success in Israel. Plant traits considered were shoot and root biomass production, root-shoot ratio, shoot height, root length, number of inflorescences, achene number and mass, and life span. Achenes (seeds) of all populations were germinated under common growing conditions to produce F1 achenes. F1 seedlings were grown in a large-scale common garden aeroponic system until flowering and then harvested. Introduced populations exhibited marked differences in measured parameters than native populations. Notably, root length of introduced populations exceeded 5 m, almost fourfold greater than that of native populations, allowing access to soil moisture and nutrients from deep sand layers and late-summer flowering.

CARBON DIOXIDE

Germany

Measuring Whole Plant Carbon Dioxide Exchange with the Environment Reveals Opposing Effects of the gin2-1 Mutation in Shoots and Roots of *Arabidopsis thaliana* (Brauner et al., 2015).

Brauner, K. et al., University of Stuttgart, Institute of Biomaterials and biomolecular Systems, Department of Plant Biotechnology, Stuttgart, Germany.

An investigation was conducted on the effect of reduced leaf glucokinase activity on plant carbon balance using a cuvette for simultaneous measurement of net photosynthesis in above ground plant organs and root respiration. The gin2-1 mutant of *Arabidopsis thaliana* is characterized by a 50% reduction of glucokinase activity in the shoot, while activity in roots is about fivefold higher and similar to wild-type plants. High levels of sucrose accumulating in leaves during the light period correlated with elevated root respiration in gin2-1. Despite substantial respiratory losses in roots, growth retardation was moderate, probably because photosynthetic carbon fixation was simultaneously elevated in gin2-1. The data indicate that futile cycling of sucrose in shoots exerts a reduction on net carbon dioxide gain, but this is over-compensated by the prevention of exaggerated root respiration resulting from high sucrose concentration in leaf tissue.

CARPETWEED

India
In Vitro Growth Profile and Comparative Leaf Anatomy of the C3–C4 Intermediate Plant *Mollugo nudicaulis* Lam (Barupal, 2018).

Barupal, Meena et al., Biotechnology Unit, Department of Botany Jai Narain Vyas University Jodhpur, India.

This paper describes a study of in vitro growth profiling of the Mollugo nudicaulis Lam., commonly known as John's folly or naked-stem carpetweed, and comparative leaf anatomy under in vitro and ex vitro conditions is an ephemeral species of tropical regions. The plant is ideal to study the eco-physiological adaptations of C3–C4 intermediate plants. In vitro propagation of the plant was carried out on Murashige and Skoog (MS) basal medium augmented with additives and solidified with 0.8% (w/v) agar-agar or 0.16% (w/v) Phytagel™. The concentration of plant growth regulators (PGRs) in the basal medium was optimized for callus induction, callus proliferation, shoot regeneration, and in vitro rooting.

CHICKPEA

Germany
Abscisic Acid Concentration, Root pH, and Anatomy do not Explain Growth Differences of Chickpea (*Cicer areitinum* L.) and Lupin (*Lupinus angustifolius* L.) on Acid and Alkaline Soils (Hartung et al., 2002).

Hartung, Wolfram et al., Universität Würzburg, Julius-von-Sachs-Institut für Biowissenschaften, Würzburg, Germany.

This paper describes a study of the anatomy of roots in aeroponic and hydroponic culture; the poor growth of narrow-leafed lupins in alkaline soil by measuring the abscisic acid (ABA) concentrations of leaves, roots, soils, and transport fluids of chickpea and lupin plants growing in alkaline and acidic soils. The paper also includes information about the root anatomy; cytoplasmic and vacuolar pH, and ABA analyses.

CHRYSANTHEMUM

Colombia
Absorption Curves—Mineral-Extraction Under an Aeroponic System for White Chrysanthemum (*Dendranthema grandiflorum* (Ramat.) Kitam. cv. Atlantis White) (Chica Toro et al., 2018).

Chica Toro, Faber de Jesús et al., Universidad Católica de Oriente, Antioquia, Colombia.

Absorption and extraction curves using an aeroponic system for White chrysanthemum cv. Atlantis White were constructed and determined during periods of maximum and minimum nutrient accumulation from above ground, root, and total biomass. Vegetative cycle of the plant, measured from the day after transplantation to aeroponic system, and the first day of court lasted 49 days. In this study, the reported elements (N, P, K, Ca, Mg, and S), its accumulation during the first stage did not exceeded 29%, except for S, which reached 30.19%. Consequently, the 70% nutrients were absorbed in the second stage of development, which coincided with the plant reproductive stage.

CORN

Germany

Chemical Composition of Apoplastic Transport Barriers in Relation to Radial Hydraulic Conductivity of Corn Roots (*Zea mays* L.) (Zimmermann et al., 2000).

Zimmermann, Hilde Monika et al., Lehrstuhl Pflanzenökologie, Universität Bayreuth, Bayreuth, Germany.

This paper describes a study of the hydraulic conductivity of roots (Lpr) of 6–8-day-old maize seedlings and its relationship to the chemical composition of apoplastic transport barriers in the endodermis and hypodermis (exodermis), and to the hydraulic conductivity of root cortical cells. Roots were cultivated in two different ways. When grown in aeroponic culture, they developed an exodermis (Casparian band in the hypodermal layer), which was missing in roots from hydroponics. The development of Casparian bands and suberin lamellae was observed by staining with berberin-aniline-blue and Sudan-III. The compositions of suberin and lignin were analyzed quantitatively and qualitatively after depolymerization (BF3/methanol-transesterification, thioacidolysis) using gas chromatography/mass spectrometry. Root Lpr was measured using the root pressure probe, and the hydraulic conductivity of cortical cells (Lp) using the cell pressure probe. Roots from the two cultivation methods differed significantly in (1) the Lpr evaluated from hydrostatic relaxations (factor of 1.5), and (2) the amounts of lignin and aliphatic suberin in the hypodermal layer of the apical root zone. Aliphatic suberin is thought to be the major reason for the hydrophobic properties of apoplastic barriers and for their relatively low permeability to water. No differences were found in the amounts of suberin in the hypodermal layers of basal root zones and in the endodermal layer. It was concluded that changes in the hydraulic conductivity of the apoplastic rather than of the cell-to-cell path were causing the observed changes in root Lpr.

COWPEA

China

Effects of Aluminum (+3) on the Biological Characteristics of Cowpea Root Border Cells (Chen et al., 2008).

Chen, Wenrong et al., School of Chemical Engineering and Technology, Tianjin University, Tianjin, China.

This team investigated root border cells (RBC) that are cells surrounding the root apex for viability, formation, and pectin methylesterase (PME) activity of the root caps during RBC development in cowpea (*Vigna ungniculata* ssp. sesquipedalis) under aeroponic culture. The results showed that the border cells formed almost synchronously with the emergence of the root tip. The number of border cells reached the maximum when roots were approximately 15 mm long. Pectin methylesterase (PME) activity of the root cap peaked at a root length of 1 mm. Root border cells separated from the root cap died within 24 h under aluminum (+3) stress while those still attached to the root cap maintained 85% viability at 48 h after treatment. The PME activity did not differ significantly under different aluminum treatments.

CRANBERRY

USA

Measurement of Short-Term Nutrient Uptake Rates in Cranberry by Aeroponics (Barak et al., 1996).

Barak, P. et al., Department of Soil Science, Univ.
of Wisconsin-Madison, Madison, USA.

The paper was focused on the determination of whether nutrient uptake rates could be calculated for aeroponic systems by difference using measurements of concentrations and volumes of input and efflux solutions. Data were collected from an experiment with cranberry plants (*Vaccinium macrocarpon* Ait. Cv. Stevens) cultured aeroponically with nutrient solutions containing, various concentrations of ammonium-N and isotopically labeled nitrate-N. Aeroponics, a soil-less plant culture in which fresh nutrient solutions are intermittently or continuously misted on to plant roots, is capable of sustaining plant growth for extended periods of time while maintaining a constantly refreshed nutrient solution. Although used relatively extensively in commercial installations and in root physiology research, use of aeroponics in nutrient studies is rare. The object of this study was to examine whether Validation of the calculated uptake rates was sought by: (1) evaluating charge balance of the solutions and total ion uptake (including proton efflux) and (2) comparison with N-isotope measurements. The results show that charge balance requirements were acceptably satisfied for individual solution analyses and for total ion uptake when proton efflux was included. Use of aeroponic systems for nondestructive measurement of water and ion uptake rates for numerous other species and nutrients appears promising.

USA

Rate of Ammonium Uptake by Cranberry (*Vaccinium macrocarpon* Ait.) Vines in the Field is Affected by Temperature (Roper et al., 2004).

Roper, T. R. et al., Department of Horticulture, University
of Wisconsin-Madison, Madison, Wisconsin, USA.

This study focused ons how quickly cranberries in the field take up fertilizer-derived ammonium nitrogen. Nitrogen fertilizer application is a universal practice among cranberry growers. Cranberries only use ammonium nitrogen sources. Ammonium sulfate labeled with 15N was applied in field locations in Oregon, Massachusetts, New Jersey, and Wisconsin. Samples of current season growth were collected daily for 7 days beginning 24 h after fertilizer application. In all cases 15N was detectable in the plants from treated plots by 24 h following application. Additional nitrogen was taken up for the next 3–5 days depending on the location. With the exception of Oregon, the maximum concentration of 15N was found by day 7. Oregon was the coolest of the sites in this research. To determine a temperature response curve for N uptake in cranberry, cranberry roots were exposed to various temperatures in aeroponics chambers while vines were at ambient greenhouse temperatures. The optimum temperature for N uptake by cranberry vines was 18°C–24°C. This research suggests that ammonium fertilizers applied by growers and irrigated into the soil (solubilized) are taken up by the plant within 1 day following application. Soil and root temperature is involved in the rate of N uptake.

CUCUMBERS

Colombia

Growing Degree Days Accumulation in a Cucumber (*Cucumis sativus* L.) Crop Grown in an Aeroponic Production Model (Hoyos et al., 2012).

Hoyos García et al., Estudiante Ingeniería Agronómica. Universidad Nacional de Colombia—Sede Medellín, Facultad de Ciencias Agrarias— Departamento de Ciencias Agronómicas. Medellín, Colombia.

This team studied variables which may affect the efficiency and crop production under an aeroponic system. It was determined that 726 and 660 Growing Degree Days (GDD) corresponding to 73 and 64 days were required for the commercial matherials Dasher II and Poinsset 76, respectively. The effect of two misting time periods of 30 and 60 s followed by a 4 min interval during the day, were evaluated over leaf area and stem and leaves dry weight, using the hybrid Dasher. No statistical significant differences were found suggesting that the 30 s time period is the best choice since it reduces electric energy costs. The effect of three different nutrient solutions: Hoagland and Arnon, Aeroponicos 100% and Aeroponicos 50%, were tested for leaf area, dry weight, fruit weight and number. The results allowed implementing variables to increase efficiency on a cucumber aeroponic crop system, some of which may improve the economic and environmental performance of cucumber crop using this technology.

ELM

USA

Vegetative Propagation of American Elm (*Ulmus americana*) Varieties from Softwood Cuttings (Oakes et al., 2012).

Oakes, Allison D. et al., Department of Plant Science and Biotechnology, State University of New York, College of Environmental Science and Forestry, Syracuse, New York, USA.

This team studied softwood cuttings of American elm varieties 'Jefferson', 'New Harmony', 'Princeton', 'R18-2', 'Valley Forge', and a tissue-cultured nontransformed control clone (BP-NT) that were rooted using three different treatments to determine which method would be most suitable for small-scale propagation. The treatments included aeroponic chambers, an intermittent-mist bench in a greenhouse, and Grodan rootplugs soaked in a nutrient solution. The rootplug treatment had the highest percentage of rooted shoots (44%) followed by the intermittent-mist bench treatment (20%) and lastly by the aeroponics chambers (10%). The rooted cuttings from the rootplug treatment also looked substantially healthier and had more fresh growth 4 weeks after potting than the other two treatments. The Grodan rootplug treatment is recommended, but additional testing can be useful to improve the overall rooting percentage.

EUCALYPTUS

Australia
Influence of Low Oxygen Levels in Aeroponics Chambers on Eucalyptus Roots Infected with *Phytophthora cinnamomi* (Burgess et al., 1998).

Burgess, T. et al., Environmental and Conservation Sciences, Murdoch University.
Perth, Australia

This study focused on the design of aeroponics root chambers to evaluate the influence of low oxygen on disease development in clones of *Eucalyptus marginata* susceptible or resistant to infection by *Phytophthora cinnamomi*. Actively growing 7-month-old clones of *E. marginata* were transferred into the aeroponics chambers, into which a nutrient solution was delivered in a fine spray, providing optimal conditions for root growth. Root extension during hypoxia was greatly reduced. Lesion development was least for roots exposed to hypoxia and greatest for roots exposed to anoxia for 6 h, suggesting increased resistance of *E. marginata* to *P. cinnamoni* following hypoxia.

Australia
Action of the Fungicide Phosphite on *Eucalyptus marginata* Inoculated with *Phytophthora cinnamomic* (Jackson et al., 2000).

T. J. Jackson et al., School of Biological Sciences, Murdoch
University, Perth, Western Australia, Australia.

This team studied the chemical mechanisms behind phosphite protection in the control of *P. cinnamomi* in *E. marginata* (jarrah). Using an aeroponics system, jarrah clones with moderate resistance to *P. cinnamomi* were treated with foliar applications of phosphite (0 and 5 g/L). The roots were inoculated with zoospores of *P. cinnamomi* at 4 days before and 0, 2, 5, 8, and 14 days after phosphite treatment. Root segments were then analyzed for activity of selected host defense enzymes [4-coumarate coenzyme A ligase (4-CL), cinnamyl alcohol dehydrogenase (CAD)] and the concentration of soluble phenolics and phosphite.

Australia

Effects of Hypoxia on Root Morphology and Lesion Development in *Eucalyptus marginata* Infected with *Phytophthora cinnamomic* (Burgess et al., 1999a).

Burgess, T. et al., School of Biological Sciences, Murdoch University, Perth, Western Australia, Australia.

This team studied plants of a *Eucalyptus marginata* clone (1JN30) by growing them in aeroponics chambers that could be sealed to allow the manipulation of oxygen levels in the root environment. Roots were grown for varying periods of hypoxia (0, 2, 5, 11, or 29 days) before being inoculated with zoospores of *Phytophthora cinnamomi*. A similar set of roots was inoculated 3 days after the hypoxia treatments. Root extension was reduced at the end of all the hypoxia treatments. Six days after the hypoxia treatments, root extension had returned to normal for roots that had been exposed to 5 days of hypoxia, while for roots exposed to 11 or 29 days, extension was half the normal rate. Longitudinal sections of root tips after 5, 11, or 29 days of hypoxia indicated that the treatment caused a reduction in cell division, but not in cell expansion. In the case of roots exposed to 2 days of hypoxia, the apical meristem appeared normal at the end of the treatment, but 3 days after the return to normal oxygen conditions many of the apical meristems had died and the roots had a clubbed appearance. Thus, *E. marginata* roots have an acclimatization period to hypoxia of between 2 and 5 days, after which they can tolerate hypoxia for extended periods.

Australia

Increased Susceptibility of *Eucalyptus marginata* to Stem Infection by *Phytophthora cinnamomi* Resulting from Root Hypoxia (Burgess et al., 1999b).

Burgess, T. et al., School of Biological Sciences, Murdoch University, Perth, Western Australia, Australia.

This team examined whether eucalyptus marginata grown on rehabilitated bauxite mines and exposed to waterlogging (hypoxia) at the roots, as well as ponding around the stems at the soil surface may predispose stems of *Eucalyptus marginata* to infection by *Phytophthora cinnamomi*. Plants of *E. marginata* clones resistant and susceptible to *P. cinnamomi* were grown in an aeroponics system that could be sealed to allow the manipulation of oxygen levels in the root zone to simulate waterlogging. Plants grown under normal oxygen conditions were compared with those whose root zone was exposed to hypoxia (2 mg O_2/L) before, during or after the stems were inoculated with zoospores of *P. cinnamomi*. Inoculation was achieved by constructing receptacles around the stems that could hold water and zoospores. The greatest difference between colonized and noninoculated plants was observed at the colonization front. Peroxidase activity increased after tissues were colonized, rather than preceding the colonization as seen with the other enzymes. The stress induced by root hypoxia remained after roots were returned to normal oxygen conditions.

EVERGREEN

India
Aeroponics for Adventitious Rhizogenesis in Evergreen Haloxeric Tree *Tamarix aphylla* (L.) Karst.: Influence of Exogenous Auxins and Cutting Type (Sharma et al., 2018).

Sharma, Udit et al., Biotechnology Unit, Department of Botany, Jai Narain Vyas University, Jodhpur, India.

This research evaluated an aeroponics technique for vegetative propagation of T. aphylla. Effect of various exogenous auxins (indole-3-acetic acid, indole-3-butyric acid, and naphthalene acetic acid) at different concentrations (0.0, 1.0, 2.0, 3.0, 5.0, and 10.0 mg/L) was examined for induction of adventitious rooting and other morphological features. Among all three auxins tested individually, maximum rooting response (79%) was observed with IBA 2.0 mg L. However, stem cuttings treated with a combination of auxins (2.0 mg L IBA and 1.0 mg 1 IAA) for 15 min resulted in 87% of rooting response. Among three types of stem cuttings (apical shoot, newly sprouted cuttings, and mature stem cuttings), maximum rooting (~90%) was observed on mature stem cuttings. A number of roots and root length were significantly higher in aeroponically rooted stem cuttings as compared to stem cuttings rooted in soil conditions. Successfully rooted and sprouted plants were transferred to polybags with 95% survival rate.

FIR

Canada
Family Variation in Nutritional and Growth Traits in Douglas-fir Seedlings (Hawkins, 2007).

Hawkins, B. J., Centre for Forest Biology, University of Victoria, Victoria, BC, Canada.

This research assessed nitrogen (N) uptake and utilization in seedlings of six full-sib families of coastal Douglas fir [*Pseudotsuga menziesii* (Mirb.) Franco] known to differ in growth rate at the whole plant and root levels. Seedlings were grown in soil or aeroponically with high and low nutrient availability. Consistent family differences in growth rate and Nutilization index were observed in both soil and aeroponic culture, and high-ranking families by these measures also had greater net N uptake in soil culture. Two of the three families found to be fast-growing in long-term field trials exhibited faster growth, higher nitrogen utilization indices and greater net nitrogen uptake at the seedling stage. Mean family net influx of ammonium (NH_4^+) and efflux of nitrate (NO_3^-) in the high- and low-nutrient treatments were significantly correlated with measures of mean family biomass. The high-nutrient availability treatment increased mean net fluxes of NH_4^+ and NO_3^- in roots. These results indicate that efficiency of nutrient uptake and utilization contribute to higher growth rates of trees.

Canada

Douglas-Fir Seedling Response to a range of Ammonium Nitrate Ratios in Aeroponic Culture (Everett et al., 2010).

Everett, Kim T. et al., Centre for Forest Biology,
University of Victoria, Victoria, BC, Canada.

This study focused on the determination of the most favorable nitrogen (N) source ratio of ammonium (NH_4^+) and nitrate (NO_3^-) for aeroponically-grown Douglas-fir when pH was maintained at pH 4.0. Seedlings were grown in controlled environments with solutions containing 0:100, 20:80, 40:60, 60:40, 80:20, or 100:0 NH_4^+:NO_3^- ratios. Nutrient additions in the aeroponic culture units were controlled by solution conductivity set points. Seedling growth and nutrient allocation was observed for 45 days. Different NH_4^+:NO_3^- ratios resulted in significant differences in the rate of N addition, growth, morphology, and nutrient allocation. Seedlings grown in solutions containing 60% or 80% NO_3^- were characterized by a combination of high growth and photosynthetic rates, high and stable internal plant N concentrations, and sufficient levels of other essential nutrients. High proportions of NH_4^+ in solution resulted in low rates of N addition, stunted lateral root growth, and may have been toxic.

FOOD SECURITY

USA

Impact of Climate Change on Food Security and Proposed Solutions for the Modern City (He et al., 2013).

He, J. et al., Dept., University of Wisconsin-Madison, Madison, USA.

This team used novel aeroponics technology to produce fresh vegetables in Singapore since 1997. This innovative system allows production of all vegetables, all year round by simply cooling the roots while the aerial parts grow under tropical ambient environments. While it is not possible for arable land to be expanded horizontally, an urban farming system could increase production area through vertical extensions. Vertical stacking of troughs is constrained by the weight factor of the troughs.

USA

Aeroponics: A Sustainable Solution for Urban Agriculture (Miller, 2020).

Matthew Miller, Environmental Law Institute, Washington, DC.

This research was conducted on urban agriculture with a focus on aeroponics. Despite their well-documented benefits, most urban farms struggle to make it in the marketplace. In 2016, the *British Food Journal* found that the average sales for all urban farms were just $54,000 per year with the median income level sitting at $5,000. Another 2010 study conducted by the *Journal of Extension* found that among 243 self-reporting urban farming projects, 49% had $10,000 or less of total gross sales. Unfortunately, for those hoping to employ members of the local community, these bottom lines simply cannot support full-time employment, let alone for a single owner or manager.

UK

Can the Optimization of Pop-up Agriculture in Remote Communities Help Feed the World? (Gwynn-Jones et al., 2018).

Gwynn-Jones, Dylan et al., IBERS, Aberystwyth University, UK.

This team explored past research and crop growth in remote areas like the polar regions on Earth, and in space, with the scope to improve on the systems used in these areas to date. They introduce biointensive agricultural systems and 3D growing environments, intercropping in hydroponics, aeroponics and the production of multiple crops from single growth systems. To reflect the flexibility and adaptability of these approaches to different environments they called this type of enclosed system 'pop-up agriculture.' The vision here is built on sustainability, maximizing yield from the smallest growing footprint, adopting the principles of a circular economy, using local resources and eliminating waste.

China

Modern Plant Cultivation Technologies in Agriculture Under Controlled Environment: A Review on Aeroponics (Lakhiar et al., 2018a).

Lakhiar, Imran Ali et al., Key Laboratory of Modern Agricultural Equipment and Technology, Ministry of Education, Institute of Agricultural Engineering, Jiangsu University, Zhenjiang, Jiangsu, China.

This team reviewed a novel approach to plant cultivation under soil-less culture. At present, global climate change is expected to raise the risk of frequent drought. Agriculture is in a phase of major change around the world and dealing with serious problems. In the future, it would be a difficult task to provide a fresh and clean food supply for the fast-growing population using traditional agriculture. Under such circumstances, soil-less cultivation is the alternative technology to adapt effectively such as a Hydroponic and Aeroponics system. In the aeroponics system, plant roots are hanging in the artificially provided plastic holder and foam material replacement of the soil under controlled conditions. The roots are allowed to dangle freely and openly in the air. However, the nutrie1nt rich-water delivered with atomization nozzles. The nozzles create a fine spray mist of different droplet size intermittently or continuously. This review concludes that aeroponics system is considered the best plant growing method for food security and sustainable development. The system has shown some promising returns in various countries and was recommended as the most efficient, useful, significant, economical and convenient plant growing system soil and other soil-less methods.

USA

The Vertical Farm: A Review of Developments and Implications for the Vertical City. (Al-Kodmany, 2018).

Al-Kodmany, Kheir, Department of Urban Planning and Policy, College of Urban Planning and Public Affairs, University of Illinois at Chicago, Chicago, IL, USA.

This is a review paper on the emerging need for vertical farms by examining issues related to food security, urban population growth, farmland shortages, "food miles", and associated greenhouse gas (GHG) emissions. Urban planners and agricultural leaders have argued that cities will need to produce food internally to respond to demand by increasing population and to avoid paralyzing congestion, harmful pollution, and unaffordable food prices. The paper examines urban agriculture as a solution to these problems by merging food production and consumption in one place, with the vertical farm being suitable for urban areas where available land is limited and expensive. Luckily, recent advances in greenhouse technologies such as hydroponics, aeroponics, and aquaponics have provided a promising future to the vertical farm concept. These high-tech systems represent a paradigm shift in farming and food production and offer suitable and efficient methods for city farming by minimizing maintenance and maximizing yield. Upon reviewing these technologies and examining project prototypes, it was found that these efforts may plant the seeds for the realization of the vertical farm. The paper, however, closes by speculating about the consequences, advantages, and disadvantages of the vertical farm's implementation. Economic feasibility, codes, regulations, and a lack of expertise remain major obstacles in the path to implementing the vertical farm.

USA

Feeding 11 billion on 0.5 billion hectare of Area under Cereal Crops (Rattan, 2016).

Lal, Rattan, School of Environment and Natural Resources, The Ohio State University, Columbus, OH, USA.

The paper evaluated the need for increased global food production. Despite impressive increases in global grain production since 1960s, there are 795 million food-insecure and ~2 billion people prone to malnutrition. Further, global population of 7.4 billion in 2016 is projected to increase to 9.7 billion by 2050, with almost all increase occurring in developing countries. Thus, it is recommended that global food production be increased by 60%–70% between 2005 and 2050. Global crop production increased threefold between 1965 and 2015 with a net increase of only 67 million ha (Mha) of cropland area. Nonetheless, agronomic yield of food staples can still be tripled or quadrupled in Sub-Saharan Africa (SSA), South Asia (SA), and the Caribbean by a widespread adoption of site-specific best management practices of sustainable intensification (SI). Rather than expanding the area under cropland, agriculturally marginal and degraded soils can be set aside for nature conservancy. The global average cereal yield of 3.27 mg/ha in 2005 can be increased to 5 mg/ha by 2050, 6 mg/ha by 2080 and 7 mg/ha by 2100 through SI of agroecosystems in SSA, SA, and elsewhere. The strategy of 'producing more from less' necessitates restoration of soil health and increasing soil organic carbon (SOC) concentration to be more than 1.5%–2.0% in the rootzone. The goal of SOC sequestration is in accord with the '4 per Thousand' initiative proposed at COP21 in 2015. Therefore, global food demands can be met despite the decreasing trends in the per capita cropland area by 2050—0.17 ha in the world and 0.15 ha in developing countries. While enhancing productivity by SI, the strategy is to simultaneously reduce food waste, increase access and distribution of food, and promote plant-based diet. The goal is to reconcile high production with better environmental quality, develop urban agriculture (aquaponics, aeroponics, and vertical farms), promote nutrition-sensitive farming, and restore degraded soils. Sustainable intensification of agroecosystems can produce enough food grains to feed one person for a year on 0.045 ha of arable land.

Russia

УРБАНИЗИРОВАННОЕ АГРОПРОИЗВОДСТВО (СИТИ-ФЕРМЕРСТВО) КАК ПЕРСПЕКТИВНОЕ НАПРАВЛЕНИЕ РАЗВИТИЯ МИРОВОГО АГРОПРОИЗВОДСТВА И СПОСОБ ПОВЫШЕНИЯ ПРОДОВОЛЬСТВЕННОЙ БЕЗОПАСНОСТИ ГОРОДОВ (Руткин et al., 2017).

Руткин, Н. М. et al., Astrakhan State Technical University, Federal State Budgetary Educational Institution, Astrakhan, Russia.

This paper focused on the outlook of the development of world urban agrotechnologies ("city-farming") by means of key innovation technological and market trends analysis. It is noted that the tendencies to reduction of the area of productive lands, exhausting ecosystem resources, including World ocean resources, harmful consequences of the climate changing are the main limiting factors of the development of traditional agriculture and supplying food products to the growing population of the world. The remote territories of mass food production from the mass markets result in a large amount of waste products (food losses) in supply chains, along with decreasing product quality and raising costs. Growth of the world population, increasing concentration of urban citizens along with changing of consumers' food preferences towards "health", "natural", "organic" food bring up the development of an additional, or alternate, system of uninterrupted supply or self-provision of cities with food products, ensuring future food security. The article highlights the prospect of developing the international branch of agriculture in terms of its transition to the high-tech stage of development ("AgTech"), and reviews the innovation technologies inseparable from that transition. It has been found that the development of the urban agrotechnologies (cityfarming), as a combination of innovative high-performance agro-practices of the food production in urban environment, can step up the level of food security due to increasing food availability in qualitative and quantitative aspects. The review of main city-farming technologies in accordance with directions of its practical applications was done for the first time. The conception "urban agrotechnologies" ("city-farming") has been defined as the scientific term.

USA
Food Security (Steen, 2016).

Steen, Hoyer J., Computational and Systems Biology Program, Washington University in St. Louis, Donald Danforth Plant Science Center, St. Louis, MO, USA.

This paper focused on research that concluded that plant–microbe interaction could potentially improve crop productivity grown in vertical indoor farms, on new light-emitting diode technologies has helped advance indoor farming, and water savings offered by aeroponic and hydroponic growth methods.

FUNGI

Taiwan
Spore Development of *Entrophospora kentinensis* in an Aeroponic System (Wu et al., 1995).

Chi-Guang Wu et al. at the Soil Microbiology Laboratory, Agricultural Chemistry Department, Taiwan Agricultural Research Institute, Taiwan, Republic of China.

This team propagated *Entrophospora kentinensis* with bahia grass and sweet potato in an aeroponic system. Spores were produced 6 weeks later after host plants were transferred to an aeroponic chamber. After spores mature, the terminal vesicles of saccules and the lower hyphal stalks degenerate and leave two scars. Differences in spore ontogeny between Acaulospora and Entrophospora are discussed.

USA
Production of Vesicular-Arbuscular Mycorrhizal Fungus Inoculum in Aeroponic Culture (Hung et al., 1988).

Hung, Ling-Ling L. et al., Soil Science Department, University of Florida, Florida.

This team grew bahia grass (*Paspalum notatum*) and industrial sweet potato (*Ipomoea batata*) colonized by *Glomus deserticola*, *G. etunicatum*, and *G. intraradices* in aeroponic cultures. After 12–14 weeks, all roots were colonized by the inoculated vesicular-arbuscular mycorrhizal fungi. Abundant vesicles and arbuscules formed in the roots, and profuse sporulation was detected intra-and extraradically. Within each fungal species, industrial sweet potato contained significantly more roots and spores per plant than bahia grass did, although the percent root colonization was similar for both hosts. Aeroponically produced *G. deserticola* and *G. etunicatum* inocula retained their infectivity after cold storage (4°C) in either sterile water or moist vermiculite for at least 4 and 9 months, respectively.

USA
Movement and Containment of Microbial Contamination in the Nutrient Mist Bioreactor (Sharaf-Eldin et al., 2006).

Mahmoud A. Sharaf-Eldin et al., Department of Biology and Biotechnology Worcester Polytechnic Institute, Worcester, USA.

The study was conducted on the movement and control of contaminants in the mist bioreactor, the spore-forming microbes Penicillium chrysogenum and Bacillus subtilis by deliberately inoculating into three possible locations in the reactor: the growth chamber (GC), the medium reservoir (R), or the mist-generating chamber (MG). Compared to inoculation into either R or MG regions, the growth of P. chrysogenum inoculated into the GC required 3 more days (c. 60% more time) to move throughout the rest of the reactor. In contrast, regardless of where B. subtilis was inoculated (GC, R, or MG), it took 7 d to contaminate the entire system. The movement of filamentous fungi and bacteria seems to follow the same route of contamination throughout this reactor.

Australia
Improved Aeroponic Culture of Inocula of Arbuscular Mycorrhizal Fungi (Mohammad et al., 2000).

Mohammad, A. et al., Department of Biological Sciences, University of Western Sydney, Australia.

This study compared conventional atomizing disc aeroponic technology with the latest ultrasonic nebulizer technology for production of *Glomus intraradices* inocula. The piezo ceramic element technology used in the ultrasonic nebulizer employs high-frequency sound to nebulize nutrient solution into microdroplets 1 μm in diameter. Growth of pre-colonized arbuscular mycorrhizal (AM) roots of Sudan grass was achieved in both chambers used but both root growth and mycorrhization were significantly faster and more extensive in the ultrasonic nebulizer system than in the atomizing disc system. Thus, the latest ultra-sonic nebulizer aeroponic technology appears to be superior and an alternative to conventional atomizing disc or spray nozzle systems for the production of high-quality AMF inocula. These can be used in small doses to produce a large response, which is a prerequisite for commercialization of AMF technology.

USA

Review: Beneficial Bacteria and Fungi in Hydroponic Systems: Types and Characteristics of Hydroponic Food Production Methods (Lee et al., 2015).

Lee, Seungjun et al., Environmental Science Graduate Program, The Ohio State University, USA.

This is a review article on information concerning hydroponic systems, including the different types and methods of operation; trends, advantages and limitations, the role of beneficial bacteria and fungi in reducing plant disease and improving plant quality and productivity. In order to produce more and improved hydroponic crops, a variety of modified hydroponic systems have been developed, such as: the wick, drip, ebb-flow, water culture, nutrient film technique, aeroponic, and windowfarm systems. According to numerous studies, hydroponics have many advantages over field culture systems, such as: reuse of water, ease in controlling external factors, and a reduction in traditional farming practices (e.g., cultivating, weeding, watering, and tilling). However, several limitations have also been identified in hydroponic culture systems, i.e., high setup cost, rapid pathogen spread, and a need for specialized management knowledge.

Belgium

Methods for large-scale production of AM fungi: past, present, and future (Ijdo et al., 2011).

Ijdo, Marleen et al., Earth and Life Institute, MycologyUniversité catholique de Louvain Louvain-la-Neuve, Belgium.

This is a review article covering the principle of in vivo and in vitro production methods that have been developed for soil- and substrate-based production techniques as well as substrate-free culture techniques (hydroponics and aeroponics) and in vitro cultivation methods for the large-scale production of AM fungi. They present the parameters that are critical for optimal production, discuss the advantages and disadvantages of the methods, and highlight their most probable sectors of application. Many different cultivation techniques and inoculum products of the plant-beneficial arbuscular mycorrhizal (AM) fungi have been developed in the last decades.

France

Inoculum of Arbuscular Mycorrhizal Fungi for Production Systems: Science Meets Business (Gianinazzi et al., 2004).

Gianinazzi, Silvio et al., INRA/Université Bourgogne Cedex, France.

The study reviewed the development of an industrial activity producing microbial inocula. It is a complex procedure that involves not only the development of the necessary biotechnological know-how but also the ability to respond to the specifically related legal, ethical, educational, and commercial requirements. At present, commercial arbuscular mycorrhizal (AM) inocula are produced in nursery plots, containers with different substrates and plants, aeroponic systems, or, more recently, in vitro. Different formulated products are available on the market, which creates the need for the establishment of standards for widely accepted quality control. One of the main tasks for both producers and researchers is to raise awareness in the public about potentials of mycorrhizal technology for sustainable plant production and soil conservation.

USA

Tissue Magnesium and Calcium Affect Arbuscular Mycorrhiza Development and Fungal Reproduction (Jarstfer et al., 1998).

Jarstfer, A. G. et al., Soil and Water Science Department,
University of Florida, Gainesville, USA.

This study was conducted on applications of high levels of magnesium sulfate in root colonization and sporulation by Glomus sp. (INVAM isolate FL329) with sweet potato and onion in aeroponic and sand culture, respectively. Onion shoot-magnesium concentrations were elevated when a nutrient solution containing 2.6 or 11.7 mM magnesium sulfate was applied. These effects on colonization and sporulation were independent of changes in tissue-P concentration. High Mg/low Ca tissue concentrations induced premature root senescence, which may have disrupted the mycorrhizal association. Their results confirm the importance of Ca for the maintenance of a functioning mycorrhiza.

India

19 Vesicular-Arbuscular Mycorrhiza: Application in Agriculture (Bagyaraj, 1992).

Bagyaraj, D.J., Department of Agricultural Microbiology,
University of Agricultural Sciences, Bangalore, India.

This is a chapter in a book entitled Methods in Microbiology and is about some of the methods used for exploiting the vesicular-arbuscular mycorrhizal symbiosis in agriculture. The first step towards application of vesicular-arbuscular mycorrhizal technology is to obtain a good starter culture. Another approach is to isolate spores of vesicular-arbuscular mycorrhizal fungi from soil by wet sieving and decanting technique. Techniques are available for the production of vesicular-arbuscular mycorrhizal inoculum in an almost sterile environment through nutrient film techniques, circulation hydroponic culture systems, aeroponic culture systems, root organ culture, and tissue culture. The chapter further explains greenhouse sanitation and mycorrhizal dependency of plants.

USA

Use of Hydrogel as a Sticking Agent and Carrier for Vesicular-Arbuscular Mycorrhizal Fungi (Hung et al., 1991).

Hung, Ling-Ling L. et al., Soil Science Department,
University of Florida, Gainesville, USA.

The study involved developing Natrosol®, a nonionic, water-soluble polymer, used as a sticking agent for direct inoculation of spores of Glomus etunicatum Becker & Gerdemann on roots of bahiagrass (*Paspalum notatum* Flugge). After 16 weeks in aeroponic culture, roots were 67%, 68%, 55%, 54%, and 41% colonized with the VAM fungus at distances 0–6, 6–9, 9–12, 12–15, and 15–18 cm below the crown, respectively. Natrosol had no effect on spore germination nor on root colonization at 14 weeks, but increased both the proportion of root length with root hairs and total root length.

USA

Beneficial Bacteria and Fungi in Hydroponic Systems: Types and Characteristics of Hydroponic Food Production Methods (Lee et al., 2015).

Lee, Seungjun et al., Environmental Science and Graduate
Program, The Ohio State University, Columbus, Ohio, USA.

This is a review of current information concerning hydroponic systems, including the different types and methods of operation; trends, advantages and limitations, the role of beneficial bacteria and fungi in reducing plant disease and improving plant quality and productivity. In order to produce more and improved hydroponic crops, a variety of modified hydroponic systems have been developed, such as: the wick, drip, ebb-flow, water culture, nutrient film technique, aeroponic, and windowfarm systems.

GRAPE

Germany

Production and Rooting behavior of rol B-Transgenic Plants of Grape Rootstock 'Richter 110' (Vitis berlandieri × V. rupestris) (Geier et al., 2008).

Thomas Geier et al., Section of Botany Geisenheim
Research Center, Geisenheim, Germany.

This study was conducted on the production and rooting behaviour of transgenic grape rootstock 'Richter 110' carrying the *Agrobacterium rhizogenes* rolB gene, which is known to promote rooting. Genetic improvement of grape rootstocks is aimed at protection against grape phylloxera and other soil-borne pests and diseases, good rooting and graft compatibility as well as adaptability to a wide range of soil and climatic conditions. Apart from the long evaluation period required, breeding is complicated by the high heterozygosity in grapes. As an alternative to traditional crossing, gene transfer permits addition of single traits, largely without affecting the genetic background of existing valuable cultivars. Rooting behaviour was examined in vitro, using tip, node and internode explants, and in aeroponic culture in the greenhouse, using single-node cuttings. Compared to internodes of nontransgenic 'Richter 110', those of rolB-transgenic clones in general showed significantly higher rooting ability and, in contrast to the former, were able to root profusely even in the absence of auxin. Cuttings of three rolB-transgenic clones in aeroponic culture produced almost twice as many primary roots as those of the nontransgenic control.

IRIS

Canada

Environmental Effects on the Maturation of the Endodermis and Multiseriate Exodermis of Iris Germanica Roots (Meyer et al., 2009).

Meyer CJ et al., Department of Biology, University of Waterloo, Waterloo, Ontario, Canada.

This team investigated the development and apoplastic permeability of Iris germanica roots with a multiseriate exodermis (MEX). The effects of different growth conditions on MEX maturation were also tested. In addition, the exodermises of eight Iris species were observed to determine whether their mature anatomy correlated with habitat.

CHINESE CABBAGE

Indonesia

Irrigation Efficiency and Uniformity of Aeroponics System a Case Study in Parung Hydroponics Farm (Prastowo, 2007).

Prastowo et al., Dept. of Agricultural Engineering, Faculty of Agricultural Technology, Bogor Agricultural University, Bogor, Indonesia.

This study evaluated the irrigation efficiency and coefficient of nonuniformity (CU) of the existing aeroponics system for one growing season of petsai (*Brassica pekinensis* L) in Parung—Bogor, Indonesia. The evaluation covers the CU of the spray discharge, pH, temperature and electrical conductivity (EC) of the nutrient solution. It was concluded that the CU of spray discharge, pH and temperature of the nutrient solution were relatively high, but the CU of EC of the nutrient solution was relatively low. Conveyance efficiency and water-use efficiency were about 84.38% and 40.09%, respectively. The average crop water requirement was about 1,457 cc/crop/season or equal to 16,870 cc/kg of petsai produced.

LETTUCE

Slovenia

Nitrate content in lettuce (*Lactuca sativa* L.) grown on aeroponics with different quantities of nitrogen in the nutrient solution [electronic resource] (Kacjan-Mar et al., 2002).

Kacjan-Mar, N. et al., Nina Kacjan Marsic Biotechnical,
University of Ljubljana, Slovenia.

This team studied the influence of different quantities of nitrogen in the nutrient solution on growth, development and nitrate content in aeroponically grown lettuce (*Lactuca sativa* L.). Three successive experiments were conducted in 1999 from April to September, in an aeroponic system. The lettuce plants, cv. Vanity, were grown in aeroponics using four different amounts of nitrogen in the nutrient solutions. The pH level was maintained between 5.5 and 6.5, and the EC between 1.8 and 2.2 mS/cm. The highest NO_3 concentration in the lettuce leaves was recorded in plants grown in nutrient solutions with the highest NO_3-N concentration (17 mM in the first, 12 mM in the second and third experiments). An acceptably low NO_3 concentration was found in the leaves of lettuce treated containing with nutrient solution 4 mM NO_3^--N in all three experiments.

USA

Growth Responses and Root Characteristics of Lettuce Grown in Aeroponics, Hydroponics, and Substrate Culture (Li et al., 2018).

Qiansheng Li et al., Department of Horticultural
Sciences, Texas A&M University, USA.

This study involved measuring the shoot and root growth, root characteristics, and mineral content of two lettuce cultivars in aeroponics compared with hydroponics and substrate culture. The results showed that aeroponics remarkably improved root growth with a significantly greater root biomass, root/shoot ratio, and greater total root length, root area, and root volume. However, the greater root growth did not lead to greater shoot growth compared with hydroponics, due to the limited availability of nutrients and water. It was concluded that aeroponics systems may be better for high value true root crop production.

Slovenia

Effects of Different Nitrogen Levels on Lettuce Growth and Nitrate Accumulation in Iceberg Lettuce (*Lactuca sativa* var. capitata L.) Grown Hydroponically under Greenhouse Conditions (Maršić et al., 2002).

Nina Kacjan Maršić et al., Institut for Fruit Growing, Viticulturae
and Vegetable Growing, University of Ljubljana, Slovenia.

This study involved the influence of different greenhouse conditions and decreasing nitrogen level in nutrient solution on growth and on nitrate accumulation and its distribution in lettuce plants. Three successive experiments were conducted on aeroponic systems in 1999. The lettuce plants cv. 'Vanity' were grown in hydroponics using 13 and 5 mM nitrate in nutrient solution. Differences among averages of fresh shoot weight measurements were statistically significant in all three aeroponic experiments.

Singapore

Effects of Root-Zone Temperature on the Root Development and Nutrient Uptake of *Lactuca sativa* L "Panama" Grown in an Aeroponic System in the Tropics (Tan et al., 2002).

Tan, Lay et al., National Science Academy Group, National Institute of Education, Nanyang Technological University, Singapore.

This team studied *Lactuca sativa* L (Panama) under conditions grown in the tropics by subjecting its roots to 20°C while its aerial portions are exposed to the hot, fluctuating temperatures in the greenhouse. This study showed that the roots were longer with a greater number of root tips and total root surface area, and smaller average root diameter as compared with those of ambient RZT (A-RZT) plants. Mineral nutrients such as nitrate, nitrogen (N), phosphorus (P), potassium (K), calcium (Ca), copper (Cu), iron (Fe), magnesium (Mg), manganese (Mn), and zinc (Zn) present in the plant shoot and root tissues were also determined. Generally, it was found that 20°C-RZT plants had higher leaf N and P concentrations on the basis of per unit dry weight compared with plants grown at A-RZT. The results also showed that total shoot and root nitrate-N, K, Ca, Cu, Fe, Mg, Mn, and Zn accumulation of 20°C-RZT plants were more than A-RZT plants.

Singapore

Growth and Photosynthetic Characteristics of Lettuce (*Lactuca sativa* L.) under Fluctuating Hot Ambient Temperatures with the Manipulation of Cool Root-Zone Temperature (Jie et al., 1998).

Jie, He et al., School of Science, National Institute of Education, Nanyang Technological University, Singapore.

The growth and photosynthetic characteristics in lettuce (*Lactuca sativa* L.) cultured in an aeroponic system at two different times of the year was studied. Midday ambient and leaf temperatures recorded in January were significantly lower than those measured in June. When the aerial parts were grown under hot ambient temperature but with their root zones exposed to 20°C, photosynthetic capacity and productivity were, respectively, about 20% and 30% higher measured from the leaves grown in January as compared with those planted in June. However, photosynthetic rate and productivity decreased by more than 50% at both periods when the whole plants were grown under hot ambient temperature as compared with those with their shoots maintained at hot ambient temperature but with their root zones exposed to a cool temperature of 20°C.

Chile

Over Fertilization Limits Lettuce Productivity Because of Osmotic Stress (Albornoz et al., 2015).

Francisco Albornoz et al., Departamento de Ciencias Vegetales, Facultad de Agronomía e Ingeniería Forestal, Pontificia Universidad Católica de Chile, Santiago, Chile.

An evaluation was conducted on the physiological response of lettuce (*Lactuca sativa* L.) to various root zone nutrient concentrations (expressed as electrical conductivity, from 0.6 to 10 dS/m), using increasing concentrations of macronutrients applied to the root zone in an aeroponic system. Leaf photosynthesis and chlorophyll fluorescence were measured using a portable infrared gas analyzer attached with a fluorometer. Leaf nutrient content was analyzed by mass spectrometry and NO_3-N was determined by flow injection analysis. Leaf photosynthetic rates increased when the solution concentration was raised from 0.6 to 4.8 dS/m, but further increases in solution concentration did not result in any differences. The enhancement in photosynthetic rates was related to higher concentrations of N, P, Mg, and S in leaves. Leaf K content was correlated with stomatal conductance. Maximum growth was achieved with solution concentrations between 1.2 and 4.8 dS/m while at 10.0 dS/m leaf production was reduced by 30%. It is concluded that at high concentration of nutrients supplied in the root zone, yield reduces because of a combination of decreased stomatal conductance and leaf area.

Chile

Effect of Different Day and Night Nutrient Solution Concentrations on Growth, Photosynthesis, and Leaf NO_3^- Content of Aeroponically Grown Lettuce (Albornoz et al., 2014).

Francisco Albornoz et al., Instituto de Investigaciones Agropecuarias, Santa Rosa, Santiago, Chile.

An evaluation of different concentrations of the nutrient solution applied during the day (D) and night (N) to aeroponically grown lettuce (*Lactuca sativa* L.) in Davis, California, USA, in the spring of 2012 with the objective of assessing the effect on growth, leaf photosynthesis, and nitrate accumulation in leaves. This study was conducted because nitrate content in leafy green vegetables has raised concerns among consumers and policy makers worldwide. Several cultural practices have been evaluated to manipulate NO_3^- content in fresh leaves with varying degrees of success.

Two different treatments in the nighttime solution concentration (D25/N75, EC: 1.8 dS/m; and D25/N50, EC: 1.2 dS/m), a day nutrient solution of EC 0.6 dS/m, plus a day and night treatment with constant EC (D50/N50, EC: 1.2 dS/m) were applied. Plant growth, leaf photosynthesis, and leaf nutrient content were evaluated after 3 weeks of growth. Switching nutrient solution concentration between day and night is a viable practice to reduce NO_3^- in lettuce leaves with no detriment to leaf production.

Korea
The Effect of LED Light Combination on the Anthocyanin Expression of Lettuce (Baek et al., 2013).

Baek, Gyeong Y. et al., Department of Bioindustrial Machinery Engineering Gyeongsang National University (Institute of Agriculture and Life Science), Jinju, Korea.

This study involved the growing of lettuce in a deep flow technique system and aeroponics method with different light combinations. The lettuce was grown in DFT system was imaged. Then image analysis and absorbance was analyzed. The Aeroponics system was imaged and the major functional elements of anthocyanin—cyanidin-3-glucoside(C3G), peonidin-3-glucoside(P3G), and delphinidin-3-glucoside(D3G)—were measured by HPLC. As a result, it turned out that in the light combination of red 53: blue 47, red 58: blue 42, the content of D3G was the highest. This study showed that blue light has significant effects on the development of anthocyanin.

Singapore
Effects of Elevated Root Zone Carbon Dioxide and Air Temperature on Photosynthetic Gas Exchange, Nitrate Uptake, and Total Reduced Nitrogen Content in Aeroponically Grown Lettuce Plants (He et al., 2010).

He, Jie et al., Natural Sciences and Science Education Academic Group, National Institute of Education, Nanyang Technological University, Singapore.

The effects of elevated root zone (RZ) carbon dioxide and air temperature on photosynthesis, productivity, nitrate ($NO(3)(-)$), and total reduced nitrogen (N) content in aeroponically grown lettuce plants was studied. Three weeks after transplanting, four different RZ carbon dioxide concentrations [ambient (360 ppm) and elevated concentrations of 2,000, 10,000, and 50,000 ppm] were imposed on plants grown at two air temperature regimes of 28°C/22°C (day/night) and 36°C/30°C. Photosynthetic CO_2 assimilation (A) and stomatal conductance (g(s)) increased with increasing photosynthetically active radiation (PAR).

Singapore
Interaction Between Potassium Concentration and Root-Zone Temperature on Growth and Photosynthesis of Temperate Lettuce Grown in the Tropics (Yi et al., 2012).

ILuo, Hong Yi et al., School of Science, National Institute of Education, Nanyang Technological University, Singapore.

Lactuca sativa L. plants at three root-zone temperatures (RZTs): 25°C, 30°C and ambient RZT (A-RZT) was grown on an aeroponic system. Three potassium (K) concentrations: −25% (minus K), control (standard K), and +25% (plus K) were supplied to plants at each RZT. Plants grown at the plus K and 25°C-RZT had the highest productivity, largest root system and highest photosynthetic capacity. The minus K plants at 25°C-RZT had the highest shoot soluble carbohydrate (SC) concentration, but they had the highest root SC concentration in the plus K plants at A-RZT. However, the highest starch concentration was found in both shoots and roots of the plus K plants at 25°C-RZT. The plus K plants had the highest shoot K concentration at 25°C-RZT, but they had the highest root K concentration at A-RZT. Highest proportion of absorbed K was partitioned to shoots when the plants were grown with the plus K at 25°C-RZT.

Japan

Dry-fog Aeroponics Affects the Root Growth of Leaf Lettuce (*Lactuca sativa* L. cv. Greenspan) by Changing the Flow Rate of Spray Fertigation (Hikosaka et al., 2015).

*Hikosaka, Yosuke et al., Department of Agricultural and
Environmental Engineering, Kobe, Japan.*

The growth characteristics and physiological activities of leaves and roots of lettuce cultivated in dry-fog aeroponics was investigated with different flow rates of nutrient dry-fog (FL, 1.0 m/s; NF, 0.1 m/s) under a controlled environment for 2 weeks and compared to lettuce cultivated using deep-flow technique (DFT). The growth of leaves of FL and DFT was not different and was significantly higher than that of NF. The amount of dry-fog particles adhering to the objects was higher in FL than in NF, so that the root growth in NF was significantly higher than that of FL. The respiration rate of roots was significantly higher in dry-fog aeroponics, but the dehydrogenase activity in the roots was significantly higher in DFT. There were no differences in the contents of chlorophyll and total soluble protein in the leaves or the specific leaf area. Photosynthetic rate and stomatal conductance were higher in dry-fog aeroponics. The contents of nitrate nitrogen, phosphate and potassium ions in the leaves were significantly higher in DFT, but the content of calcium ions was significantly higher in FL. Thus, changing the flow rate of the dry-fog in the rhizosphere can affect the growth and physiological activities of leaves and roots.

USA

Rooting for Lettuce: Aero-Green Technology, Singapore: Growing Vegetables Aeroponically—or Without Soil (Dolven, 1998).

*Dolven, Ben Congressional Research Service Specialist in Asian
Affairs, Library of Congress Washington, DC, USA.*

This is a review of the Asian Innovation Awards highlighting one of Singapore's remote northern corners, C.K. Eng's company, Aero-Green Technology, that grows lettuce, kailan, bok choi, tomatoes, and cucumbers aeroponically on beds of styrofoam with their roots dangling through holes into open air. The roots are exposed to a steady stream of nutrient-laden mist, and workers can control the air temperature around the roots and the concentration of nutrients in the water spray. According to Eng, he can grow a head of lettuce 30%–40% faster that a normal farm can, and the process uses approximately 10% of the water used by a hydroponic farm. Details of research into the aeroponic system, Eng's products, and Aero-Green's predicted profitability are provided.

LOTUS

Germany

Diurnal Variations was Conducted in Hydraulic Conductivity and Root Pressure can be Correlated with the Expression of Putative Aquaporins in the Roots of *Lotus japonicus* (Henzler et al., 1999).

Henzler, Tobias et al., Lehrstuhl Pflanzenokologie,
Universitat Bayreuth, Bayreuth, Germany.

The hydraulic conductivity of excised roots (Lpr) of the legume *Lotus japonicus* (Regel) K. Larsen grown in mist (aeroponic) and sand cultures was measured and found to vary over a 5-fold range during a day/night cycle. This behaviour was seen when Lpr was measured in roots exuding, either under root pressure (osmotic driving force), or under an applied hydrostatic pressure of 0.4 MPa which produced a rate of water flow similar to that in a transpiring plant. A similar daily pattern of variation was seen in plants grown in natural daylight or in controlled-environment rooms, in plants transpiring at ambient rates or at greatly reduced rates, and in plants grown in either aeroponic or sand culture.

Sweden/Germany

Allene Oxide Synthase, Allene Oxide Cyclase, and Jasmonic Acid Levels in *Lotus japonicus* Nodules (Zdyb et al., 2018).

Zdyb, Anna et al., Department of Ecology, Environment and
Plant Sciences, Stockholm University, Stockholm, Sweden, Georg-
August-University, Albrecht von Haller Institute for Plant Sciences,
Department of Plant Biochemistry, Göttingen, Germany.

The gene families of two committed enzymes of the jasmonic acid (JA) biosynthetic pathway, allene oxide synthase (AOS) and allene oxide cyclase (AOC), were characterized in the determinate nodule-forming model legume *Lotus japonicus*.

Jasmonic acid (JA), its derivatives and its precursor cis-12-oxo phytodienoic acid (OPDA) form a group of phytohormones, the jasmonates, representing signal molecules involved in plant stress responses, in the defense against pathogens as well as in development. Elevated levels of JA have been shown to play a role in arbuscular mycorrhiza and in the induction of nitrogen-fixing root nodules. JA levels were analyzed in the course of nodulation. Since in all *L. japonicus* organs examined, JA levels increased upon mechanical disturbance and wounding, an aeroponic culture system was established to allow for a quick harvest, followed by the analysis of jasmonic acid (JA) levels in whole root and shoot systems. Nodulated plants were compared with non-nodulated plants grown on nitrate or ammonium as N source, respectively, over a 5-week-period. JA levels turned out to be more or less stable independently of the growth conditions. However, *L. japonicus* nodules formed on aeroponically grown plants often showed patches of cells with reduced bacteroid density, presumably a stress symptom. Immunolocalization using a heterologous antibody showed that the vascular systems of these nodules also seemed to contain less AOC protein than those of nodules of plants grown in perlite/vermiculite.

MAIZE

India
Hydro and Aeroponic Technique for Rapid Drought Tolerance Screening in Maize (*Zea mays*) (Kumar et al., 2016).

Kumar, Bhupender et al., Molecular Cytogenetics and Tissue Culture Lab., Department of Crop Improvement, CSK Himachal Pradesh Agricultural University, Palampur, India.

A rapid screening technique was developed to identify drought tolerant maize (*Zea mays* L.) genotypes under field and controlled conditions at New Delhi in 2015 and 2016. The genotypes have shown variable wilting symptoms and recovery during the stress and while reverting back to hydroponic, respectively. The new rapid method could identify and verify the drought tolerant and susceptible genotypes very effectively at seedling stage and therefore it can be utilized in breeding programme for preliminary identification of drought tolerant and susceptible genotypes.

France
Relationship Between Root Structure and Root Cadmium Uptake in Maize (Redjala et al., 2011).

Redjala, Tanegmart et al., Nancy Université, INRA, Laboratoire Sols et Environnement, Vandœuvre-lès-Nancy Cedex, France.

Hypotheses were tested that (1) the cadmium (Cd) uptake is higher for maize roots grown in hydroponics than for those grown in aeroponics, (2) this difference is due to the fact that in aeroponics, root apoplastic barriers are developed more extensively than in hydroponics, and (3) the structure of maize roots grown in aeroponics is closer to the structure of roots grown in soil. A clear description of the mechanism of root cadmium absorption is required in order to understand how this toxic metal is phytoaccumulated. Apoplastic and symplastic cadmium uptake was measured by exposing the roots to a radio-labeled Cd solution and by the physical fractionation of the metal in the roots. The results obtained support the initial hypotheses. Since the characteristics of maize plants roots cultivated in aeroponics were much closer to those cultivated in soil, their kinetic parameters may be considered to be more representative when measuring uptake than those of hydroponically grown plants.

Belgium
Short-Term Control of Maize Cell and Root Water Permeability Through Plasma Membrane Aquaporin Isoforms (Hachez et al., 2012).

Hachez, Charles et al., Institut des Sciences de la Vie, Université catholique de Louvain, Croix du Sud Louvain-la-Neuve, Belgium.

The role of specific isoforms in the regulation of root water uptake was studied. The mRNA expression and protein level of specific plasma membrane intrinsic proteins (PIPs) were analyzed in *Zea mays* in relation to cell and root hydraulic conductivity. Plants were analyzed during the day/night period, under different growth conditions (aeroponics/hydroponics) and in response to short-term osmotic stress applied through polyethylene glycol (PEG). Higher protein levels during the day coincided with a higher water permeability of root cortex cells during the day compared with night period. When PEG was added to the root medium (2–8 h), cell water permeability in roots increased. These data support a role of specific isoforms in regulating root water uptake and cortex cell hydraulic conductivity in maize.

Germany

Apoplastic Transport Across Young Maize Roots: Effect of the Exodermis (Zimmermann et al., 1998).

Zimmermann, Hilde et al., Ernst Lehrstuhl Pflanzenökologie,
Universität Bayreuth, Universitätsstrasse Bayreuth, Germany.

The uptake of water and the fluorescent apoplastic dye PTS (trisodium 3-hydroxy-5,8,10-pyrenetrisulfonate) was studied by root systems of young maize (*Zea mays* L.) seedlings (age: 11–21 days) with plants which either developed an exodermis (Casparian band in the hypodermis) or were lacking it. Steady-state techniques were used to measure water uptake across excised roots. Either hydrostatic or osmotic pressure gradients were applied to induce water flows. Roots without an exodermis were obtained from plants grown in hydroponic culture. Roots which developed an exodermis were obtained using an aeroponic (mist) cultivation method. The results indicate that the radial apoplastic flows of water and PTS across the root were affected differently by apoplastic barriers (Casparian bands) in the exodermis. It is concluded that, unlike water, the apoplastic flow of PTS is rate-limited at the endodermis rather than at the exodermis. The use of PTS as a tracer for apoplastic water should be abandoned.

Germany

Pathogenicity of *Fusarium graminearum* Isolates on Maize (*Zea mays* L.) Cultivars and Relation with Deoxynivalenol and Ergosterol Contents (Asran et al., 2003).

M. R. Asran et al., Hohenheim University, Stuttgart, Baden-Württemberg.

Germany seedling blight and root rot caused by *Fusarium graminearum* isolates using an aeroponics system was determined. *F. graminearum* is an important pathogen of maize and causes seed rot and seedling blight as well as root rot, stalk rot and ear rot. In growth chamber experiments, inoculation of corn cv. 'Loyal' seeds with six different *F. graminearum* isolates reduced emergence of germlings and caused seedling death of varying degrees. This system allows nondestructive, repetitive sampling of seedlings for assessing disease progress and seedling growth. All isolates tested were able to produce deoxynivalenol (DON) in infected seedling tissue.

There was a close relationship between the degree of disease severity and DON concentration. On the other hand, a relation between disease severity and ergosterol content in the infected seedling tissues could not be detected.

France

Length of the Apical Unbranched Zone of Maize Axile Roots: Its Relationship to Root Elongation Rate (Pellerin et al., 1995).

Pellerin, Sylvain et al., INRA, Laboratoire d'Agronomie, Colmar, France.

The length of the apical unbranched zone in maize axile roots was studied. Plants were grown in an aeroponic growth chamber allowing direct measurements on individual axile roots. The total length of the roots and the length of the apical unbranched zone were measured regularly. A commonly accepted hypothesis, according to which laterals emerge at a constant distance behind the root tip, was refuted. Conversely, a linear relationship was found between the length of the apical unbranched zone and the root elongation rate. This suggests that laterals emerge on a root segment at a constant time interval after lateral primordia are differentiated.

Canada

Sites of Entry of Water into the Symplast of Maize Roots (Varney et al., 1993).

Varney, G. T. et al., Biology Department, Carleton University, Ottawa, Canada.

A new method of calculating rates of water uptake by roots from measurements of the rate of accumulation on the roots of a marker solute was developed. The paper describes the sites of accumulation of the solute, which indicate the sites where the water entered the symplast. Sulphorhodamine G (SR) was supplied in aeroponic mist culture to large maize plants with fully developed root systems. Root samples were collected after 4–8 h of transpiration in the dye-mist from both axes and branches of the main roots, and from nontranspiring (detopped) controls, frozen rapidly, freeze-substituted, and embedded and sectioned by an anhydrous procedure that preserves the SR in place. Whole mounts and sections were examined by bright-field, polarizing and epifluorescence microscopy. Major accumulations of SR were all at the outer surface of the roots, on epidermal or root hair cell walls, or, in older roots where the epidermal cells were separating or dead, on the outer wall of the hypodermis.

Germany

Protein Dynamics in Young Maize Root Hairs in Response to Macro- and Micronutrient Deprivation (Li et al., 2015).

Zhi Li et al., Department of Plant Systems Biology, University of Hohenheim, Stuttgart, Germany.

Protein abundance adjustments in 4 day old root hairs grown in aeroponic culture in the presence and absence of several macro- and micronutrients using a label-free quantitative proteomics approach was studied. Plants increase their root surface with root hairs to improve the acquisition of nutrients from the soil. In young maize seedlings, roots are densely covered with root hairs, although nutrient reserves in the seed are sufficient to support seedling growth rates for a few days. Compared to the proteome of root hairs developed under full nutrition, protein abundance changes were observed in pathways related to macronutrient (N, P, K, and Mg) deficiencies. For example, lack of N in the medium repressed the primary N metabolism pathway, increased amino acid synthesis, but repressed their degradation, and affected the primary carbon metabolism, such as glycolysis. Glycolysis was similarly affected by K and P deprivation, but the glycolytic pathway was negatively regulated by the absence of the micronutrients Fe and Zn. In contrast, the deprivation of Mn had almost no affect on the root hair proteome. Our results indicate either that the metabolism of very young root hairs adjusts to cellular nutrient deficiencies that have been already experienced or that root hairs sense the external lack of specific nutrients in the nutrient solution and adjust their metabolism accordingly.

Germany
Apoplastic Transport of Abscisic Acid through Roots of Maize: Effect of the Exodermis (Freundl et al., 1998).

Freundl, Elenor et al., Julius-von-Sachs-Institut für Biowissenschaften der Universität Würzburg, Germany.

The exodermal layers that are formed in maize roots during aeroponic culture with respect to the radial transport of cis-abscisic acid (ABA) were investigated. The decrease in root hydraulic conductivity (Lpr) of aeroponically grown roots was stimulated 1.5-fold by ABA (500 nM), reaching Lpr values of roots lacking an exodermis. Similar to water, the radial flow of ABA through roots (JABA) and ABA uptake into root tissue were reduced by a factor of about three as a result of the existence of an exodermis. Thus, due to the cooperation between water and solute transport the development of the ABA signal in the xylem was not affected. This resulted in unchanged reflection coefficients for roots grown hydroponically and aeroponically.

Philippines
Novel Temporal, Fine-Scale, and Growth Variation Phenotypes in Roots of Adult-Stage Maize (*Zea mays* L.) in Response to Low Nitrogen Stress (Gaudin et al., 2011).

Gaudin, Amélie C. M. et al., Crop Environmental Sciences Division, International Rice Research Institute, Manila, Philippines.

Root traits associated with acclimation to nutrient stress were studied. Large root systems, such as in adult maize, have proven difficult to be phenotyped comprehensively and over time, causing target traits to be missed. These challenges were overcome using aeroponics, a system where roots grow in the air misted with a nutrient solution. Applying an agriculturally relevant degree of low nitrogen (LN) stress, 30-day-old plants responded by increasing lengths of individual crown roots

(CRs) by 63%, compensated by a 40% decline in CR number. LN increased the CR elongation rate rather than lengthening the duration of CR growth. Only younger CR were significantly responsive to LN stress, a novel finding. Large-scale analysis of root hairs (RHs) showed that LN decreased RH length and density. Time-course experiments suggested the RH responses may be indirect consequences of decreased biomass/demand under LN. These results identify novel root traits for genetic dissection.

Philippines

The Effect of Altered Dosage of a Mutant Allele of Teosinte Branched 1 (tb1-ref) on the Root System of Modern Maize (Gaudin et al., 2014).

Gaudin, Amélie C. M. et al., Crop Environmental Sciences Division, International Rice Research Institute, Manila, Philippines.

Aeroponics were studied to phenotype the effects of tb1-ref copy number on maize roots at macro-, meso-, and micro scales of development. Their results consisted of: (1) an increase in crown root number due to the cumulative initiation of crown roots from successive tillers; (2) higher density of first and second order lateral roots; and (3) reduced average lateral root length. It was concluded that a decrease in Teosinte Branched 1 (Tb1) function in maize results in a larger root system, due to an increase in the number of crown roots and lateral roots. Given that decreased TB1 expression results in a more highly branched and larger shoot, the impact of TB1 below ground may be direct or indirect.

USA

Evaluation of an Aeroponics System to Screen Maize Genotypes for Resistance to *Fusarium graminearum* Seedling Blight (du Toit et al., 1997).

du Toit, Lindsey J. et al., Department of Plant Pathology, University of Illinois at Urbana-Champaign, Urbana, Illinois, USA.

A noncirculating aeroponics system as a method for rapid screening of maize genotypes for resistance to *Fusarium graminearum* seedling blight/root rot was evaluated. The system allows for nondestructive, repetitive sampling of seedlings for assessing disease progress and seedling growth. Shoot growth and root rot were assessed at 3-day intervals, and final shoot and root dry weight were determined 15 days after inoculation. The nine hybrids screened differed in severity of root rot as early as 6 days after inoculation, indicating differences in resistance to *F. graminearum*. Inoculation ` growth, root dry weight, or shoot dry weight, but differences in these agronomic traits were observed among hybrids. LH119 × LH51 and Pioneer Brand 3379 showed the greatest resistance to root rot. Area under-disease progress curve and a critical stage of disease assessment (9 days after inoculation) gave similar rankings of hybrids for root rot resistance, indicating that a single disease assessment (versus multiple assessments) may be adequate in screening for resistance with this aeroponics system.

MEDICINAL

Romania

A Study of the Cultivation of Medicinal Plants in Hydroponic and Aeroponic Technologies in a Protected Environment (Giurgiu et al., 2017).

Giurgiu, R. M. et al., University of Agricultural Science and Veterinary Medicine Cluj-Napoca, România.

Three species of medicinal plants in four hydroponic systems and one aeroponic system were studied and compared the results with plants cultivated in soil, in the same environmental conditions. Temperature, humidity, and irrigation intervals were manipulated gradually until the parameters were considered stress factors for the plants, as stated in the scientific literature. From the three plants studied, St. John's Wort showed the best results and it had a shorter time, 30 days, to the harvest peak, compared to plants cultivated in soil.

USA

Potential for Greenhouse Aeroponic Cultivation of *Urtica dioica* (Pagliarulo et al., 2004).

Pagliarulo, C.L. et al., University of Arizona Controlled Environment Agriculture Program, USA.

The applicability of aeroponic technology was determined for the cultivation of the traditionally field grown herbaceous medicinal plant *Urtica dioica*. In addition, they investigated if control of nutrient delivery and repeated harvesting practices could be utilized to increase the direct yield of desired plant parts. Comparison of root and shoot dry weights between treatments revealed: (1) *U. dioica* cultivated in soil-less medium yielded equal shoot biomass and greater root biomass than aeroponically cultivated plants; (2) potassium and phosphorus ratios within the nutrient solution had no significant impact on yield or biomass allocation; and (3) multiple harvesting of aeroponic roots and shoots yielded greater total biomass of both roots and shoots than a multi-crop replanting strategy. The results suggest aeroponic technology could be a powerful tool for the cultivation *U. dioica* as well as a variety of other important herbaceous medicinal plants. However, further optimization of the plant growing environment is required to maximize and direct growth.

India

Evaluation of Aeroponics for Clonal Propagation of *Caralluma edulis*, *Leptadenia reticulata*, and *Tylophora indica*—Three Threatened Medicinal Asclepiads (Mehandru et al., 2014).

Mehandru, Pooja et al., Biotechnology Centre, Department of Botany Jai Narain Vyas University, Jodhpur, India.

The potential of an aeroponic system for clonal propagation of *Caralluma edulis* (Paimpa) a rare, threatened and endemic edible species, *Leptadenia reticulata* (Jeewanti), a threatened liana used as promoter of health and *Tylophora indica* (Burm.f.) Merill, a valuable medicinal climber was studied. Experiments were conducted to assess the effect of exogenous auxin (naphthalene acetic acid, indole-3-butyric acid, indole-3-acetic acid) and auxin concentrations (0.0, 0.5, 1, 2, 3, 4, or 5 gL) on various root morphological traits of cuttings in the aeroponic chamber. Amongst all the auxins tested, significant effects on the length, number and percentage of rooting was observed in IBA treated nodal cuttings. All the plants sprouted and rooted aeroponically survived on transfer to soil. This is the first report of clonal propagation in an aeroponic system for these plants. This study suggests aeroponics as an economic method for rapid root induction and clonal propagation of these three endangered and medicinally important plants which require focused efforts on conservation and sustainable utilization.

USA

Cytotoxic and Other Withanolides from Aeroponically Grown *Physalis philadelphica* (Xu et al., 2018).

Xu, Ya-Ming et al., Natural Products Center, School of Natural Resources and the Environment, College of Agriculture and Life Sciences, The University of Arizona, Tucson, AZ, USA.

Eleven withanolides including six previously undescribed compounds, 16β-hydroxyixocarpanolide, 24,25-dihydroexodeconolide C, 16,17-dehydro-24-epi-dioscorolide A, 17-epi-philadelphicalactone A, 16-deoxyphiladelphicalactone C, and 4-deoxyixocarpalactone A from aeroponically grown *Physalis philadelphica* were isolated. Structures of these withanolides were elucidated by the analysis of their spectroscopic (HRMS, 1D and 2D NMR, ECD) data and comparison with published data for related withanolides. Cytotoxic activity of all isolated compounds was evaluated against a panel of five human tumor cell lines (LNCaP, ACHN, UO-31, M14 and SK-MEL-28), and normal (HFF) cells. Of these, 17-epi-philadelphicalactone A, withaphysacarpin, philadelphicalactone C, and ixocarpalactone A exhibited cytotoxicity against ACHN, UO-31, M14 and SK-MEL-28, but showed no toxicity to HFF cells.

USA

Aeroponic and Hydroponic Systems for Medicinal Herb, Rhizome, and Root Crops (Hayden, 2006).

Hayden, Anita L., Native American Botanics Corporation, Tucson, AZ, USA.

Crop production systems using perlite hydroponics, nutrient film technique (NFT), ebb and flow, and aeroponics for various root, rhizome, and herb leaf crops were studied. Hydroponic and aeroponic production of medicinal crops in controlled environments provides opportunities for improving quality, purity, consistency, bioactivity, and biomass production on a commercial scale. Ideally, the goal is to optimize the environment and systems to maximize all five characteristics. Biomass data comparing aeroponic vs. soil-less culture or field grown production of burdock root (Arctium lappa), stinging nettles herb and rhizome (*Urtica dioica*), and yerba mansa root and rhizome (*Anemopsis californica*) were presented, as well as smaller scale projects observing ginger rhizome (*Zingiber officinale*) and skullcap herb (*Scutellaria lateriflora*). Phytochemical concentration of marker compounds for burdock and yerba mansa in different growing systems were presented.

USA

Unusual Withanolides from Aeroponically Grown *Withania somnifera* (Xu et al., 2011).

Xu, Ya-ming et al., Southwest Center for Natural Products Research and Commercialization, School of Natural Resources and the Environment, College of Agriculture and Life Sciences, The University of Arizona, Tucson, AZ, USA.

The effect of growing the medicinal plant *Withania somnifera* under soil-less aeroponic conditions on its ability to produce withaferin A and withanolides was investigated. It resulted in the isolation and characterization of two compounds, 3α-(uracil-1-yl)-2,3-dihydrowithaferin A (1) and 3β-(adenin-9-yl)-2,3-dihydrowithaferin A (2), in addition to 10 known withanolides including 2,3-dihydrowithaferin A-3β-O-sulfate. 3β-O-Butyl-2,3-dihydrowithaferin A (3), presumably an artifact formed from withaferin A during the isolation process was also encountered. Reaction of withaferin A with uracil afforded 1 and its epimer, 3β-(uracil-1-yl)-2,3-dihydrowithaferin A (4). The structures of these compounds were elucidated on the basis of their high resolution mass and NMR spectroscopic data.

USA

2,3-Dihydrowithaferin A-3β-O-Sulfate, A New Potential Prodrug of Withaferin A from Aeroponically Grown *Withania somnifera* (Xu et al., 2009).

Xu, Ya-ming et al., Southwest Center for Natural Products Research and Commercialization, Office of Arid Lands Studies, College of Agriculture and Life Sciences, The University of Arizona, Tucson, AZ, USA.

The discovery of a new prodrug was reported that was produced by a medicinal plant *Withania somnifera* (L.) Dunal commonly called ashwagandha when cultured using an aeroponic technique. The new prodrug was the natural product, 2,3-dihydrowithaferin A-3β-O-sulfate (1), as the predominant constituent of methanolic extracts prepared from aerial tissues. Preparations of the roots of the medicinal plant *Withania somnifera* (L.) Dunal commonly called ashwagandha have been used for millennia in the Ayurvedic medical tradition of India as a general tonic to relieve stress and enhance health, especially in the elderly. In modern times, ashwagandha has been shown to possess intriguing antiangiogenic and anticancer activity, largely attributable to the presence of the steroidal lactone withaferin A as the major constituent.

USA

17β-Hydroxy-18-Acetoxywithanolides from Aeroponically Grown *Physalis crassifolia* and Their Potent and Selective Cytotoxicity for Prostate Cancer Cells (Xu et al., 2016).

Ya-ming Xu et al., Natural Products Center, School of Natural Resources and the Environment, College of Agriculture and Life Sciences, University of Arizona, Tucson, Arizona, USA.

Withanolides 1–11 and 16 from *Physalis crassifolia* that produced 11 new withanolides (1–11) and seven known withanolides (12–18) including those obtained from the wild-crafted plant using aeroponic growth conditions for their potential anticancer activity using five tumor cell lines were evaluated. The structures of the new withanolides were elucidated by the application of spectroscopic techniques and the known withanolides were identified by comparison of their spectroscopic data. Of these, the 17β-hydroxy-18-acetoxywithanolides 1, 2, 6, 7, and 16 showed potent antiproliferative activity, with some having selectivity for prostate adenocarcinoma (LNCaP and PC-3M) compared to the breast adenocarcinoma (MCF-7), non–small-cell lung cancer (NCI-H460), and CNS glioma (SF-268) cell lines used. The cytotoxicity data obtained for 12–15, 17, and 19 have provided additional structure-activity relationship information for the 17β-hydroxy-18-acetoxywithanolides.

USA

Biomass Production and Withaferin A Synthesis by *Withania somnifera* Grown in Aeroponics and Hydroponics (von Bieberstein et al., 2014).

von Bieberstein, Philipp et al., Southwest Center for Natural Products Research and Commercialization, School of Natural Resources & Environment, College of Agriculture and Life Sciences, University of Arizona, Tucson, AZ, USA.

The synthesis of withaferin A found in the medicinal herb *Withania somnifera* (L.) Dunal (Solanaceae) was studied. This plant was grown in two soil-less systems to determine optimal conditions for production of biomass and withaferin A, the major secondary metabolite responsible for its claimed medicinal properties. Withaferin A content was analyzed using high-performance liquid chromatography (HPLC). The results show that there was no statistically significant difference ($P > 0.05$; t test) in biomass production between the plants grown aeroponically and hydroponically. Aeroponically grown plants produced an average of 49.8 g dried aerial plant material (DW) (sd 20.7) per plant, whereas hydroponically grown plants produced an average of 57.6 g W (sd 16.0). In contrast, withaferin A content was statistically higher in plants grown hydroponically. These plants contained an average of 7.8 mg/g DW (sd 0.3), whereas the aeroponically grown plants contained an average of 5.9 mg/g DW (sd 0.6). These results demonstrate that hydroponic techniques are optimal in reproducibly and efficiently generating withaferin A. These findings may be of importance to the natural products industry in seeking to maximize production of biologically active compounds from medicinal plants.

UK

Root Phenomics of Crops: Opportunities and Challenges (Gregory et al., 2009).

Peter J. Gregory et al., Department of Soil Science, School of Human and Environmental Sciences, The University of Reading, Whiteknights, Reading, UK.

Small root systems grown in solid media using X-ray microtomography 3D noninvasive technique were measured. Reliable techniques for screening large numbers of plants for root traits are still being developed, but include aeroponic, hydroponic and agar plate systems. Coupled with digital cameras and image analysis software, these systems permit the rapid measurement of root numbers, length, and diameter in moderate (typically <1,000) numbers of plants. Usually such systems are used with relatively small seedlings, and information is recorded in 2D. However, because of the time taken to scan samples, only a small number can be screened (typically <10 per day, not including analysis time of the large spatial datasets generated) and, depending on sample size, limited resolution may mean that fine roots remain unresolved. Developments in instruments and software mean that a combination of high-throughput simple screens and more in-depth examination of root–soil interactions is becoming viable.

France

From Bioreactor to Entire Plants: Development of Production Systems for Secondary Metabolites (Nguyen et al., 2013).

Nguyen, Thi Khieu Oanh et al., Plant Biology & Innovation, Université de Picardie Jules Verne, Faculty of Sciences, Ilot des poulies, Amiens, France.

The production of tropane alkaloids from Datura innoxia and furanocoumarins from Ruta graveolens using hydroponics and aeroponics were studied. These techniques situated in-between field and fermentor scales, enable the entire plants to be used as efficient bioreactors. The production of secondary metabolites, and more specifically alkaloids, from medicinal plants is still an important objective for many research programs. When natural lead compounds have been discovered and when chemical synthesis cannot be easily performed, the extraction and purification of biomolecules from entire plants is generally the preferred solution. However, it is now established that plant cells and tissue cultures in bioreactors can constitute an alternative solution to this agronomical approach. Revisiting scientific advances made in the past decades, the ethical, legal, biological, and technological aspects are discussed in the light of the most recent literature, in order to establish a roadmap for further developments of plant secondary metabolite production systems.

MUSKMELON

China

Effects of Elevated Rhizosphere Carbon Dioxide Concentration on the Photosynthetic Characteristics, Yield, and Quality of Muskmelon (Liu et al., 2013).

Liu Yi-ling et al., College of Horticulture, Shenyang Agricultural University, Shenyang, China.

The effects of elevated rhizosphere carbon dioxide concentration on the leaf photo-synthesis and the fruit yield and quality of muskmelon during its anthesis-fruiting period using an aeroponics culture system was studied. In the fruit development period of muskmelon, as compared with those in the control (350 μL carbon dioxide/L), the leaf chlorophyll content, net photosynthetic rate (P_n), stomatal conductance (G_s), intercellular carbon dioxide concentration (C_i), and the maximal photochemical efficiency of PSII(Fv/Fm) in treatments 2,500 and 5,000 μL carbon dioxide/L decreased to some extents, but the stomatal limitation value (Ls) increased significantly, and the variation amplitudes were larger in treatment 5,000 μL CO_2/L than in treatment 2,500 μL carbon dioxide/L. Under the effects of elevated rhizosphere carbon dioxide concentration, the fruit yield per plant and the V_c and soluble sugar contents in fruits decreased markedly, while the fruit organic acid content was in adverse. It was suggested that when the rhizosphere carbon dioxide concentration of muskmelon during its anthesis-fruiting period reached to 2,500 μL/L, the leaf photosynthesis and fruit development of muskmelon would be depressed obviously, which would result in the decrease of fruit yield and quality of muskmelon.

China

Impacts of Root-Zone Hypoxia Stress on Muskmelon Growth, its Root Respiratory Metabolism, and Antioxidative Enzyme Activities (Liu et al., 2010).

Liu Yi-ling et al., College of Horticulture, Shenyang Agricultural University, Shenyang, China.

The impacts of root-zone hypoxia(10% and 5% oxygen)stress on the plant growth, root respiratory metabolism, and antioxidative enzyme activities of muskmelon at its fruit development stage using an aeroponics culture system was studied. Root-zone hypoxia stress inhibited the plant growth of muskmelon, resulting in the decrease of plant height, root length, and fresh and dry biomass. Comparing with the control (21% oxygen), hypoxia stress reduced the root respiration rate and malate dehydrogenase (MDH) activity significantly, and the impact of 5% oxygen stress was more serious than that of 10% oxygen stress. Under hypoxic conditions, the lactate dehydrogenase (LDH), alcohol dehydrogenase (ADH), pyruvate decarboxylase (PDC), superoxide dismutase (SOD), peroxidase (POD), and catalase (CAT) activities and the malondi-aldehyde (MDA) content were significantly higher than the control. The increment of antioxidative enzyme activities under 10% oxygen stress was significantly higher than that under 5% oxygen stress, while the MDA content was higher under 5% oxygen stress than under 10% oxygen stress, suggesting that when the root-zone oxygen concentration was below 10%, the aerobic respiration of muskmelon at its fruit development stage was obviously inhibited while the anaerobic respiration was accelerated, and the root antioxidative enzymes induced defense reaction. With the increasing duration of hypoxic stress, the lipid peroxidation would be aggravated, resulting in the damages on muskmelon roots, inhibition of plant growth, and decrease of fruit yield and quality.

NUTRIENTS

China
Responses of *Polygonatum odoratum* Seedlings in Aeroponic Culture to Treatments of Different Ammonium: Nitrate Ratios (Zou et al., 2017).

Zou, Tingting et al., Collaborative Innovation Center for Field Weeds Control of Hunan Province, Hunan University of Humanities, Science and Technology, Loudi, China.

The most favorable nitrogen (N) source ratio of ammonium (NH_4^+) to nitrate (NO_3^-) for aeroponic culture of Chinese fragrant solomonseal Polygonatum odoratum(Mill.) seedlings were determined. Seedlings were cultured with solutions based on 50% Hoagland formula containing 0:100, 10:90, 20:80, and 30:70 NH_4^+:NO_3^- ratios for 21 days. Activities of anti-oxidant enzymes and glutathione contents of leaves with treatments of 10:90 and 20:80 NH_4^+:NO_3^- ratios were higher than that of all-nitrate treatment, and malondialdehyde (MDA) concentrations were lower than that of all-nitrate treatment. These results supported that moderate proportion of 20% NH_4^+ in the solution provided optimal growth condition for (P)P. odoratumaeroponic culture.

Germany
The Influence of Lead Ions on the Productivity and Content of Minerals in *Phaseolus vulgaris* L. in Hydroponics and Aeroponics (Engenhart, 1984).

Engenhart, Manfred, Department of Ecology, University of Bielefeld, Germany.

The effect of lead ions on productivity and content of minerals of *Phaseolus vulgaris* L. was investigated. Plants were grown in hydroponics and aeroponics systems. The water supply under aeroponic conditions was lower than with hydroponics. Aeroponic plants treated with 5 ppm lead showed significant decrease in productivity. Especially the growth of roots was negatively influenced by lead. However, addition of 10 ppm lead in hydroponics had little effect on productivity when compared to control. The influence of lead on the mineral distribution was investigated by analysing water soluble potassium, calcium, magnesium and phosphate in leaves and roots. Translocation of calcium and magnesium from the roots to the leaves was reduced in lead treated plants in hydroponics as well as in aeroponics.

China
Experimental Study of Ultrasonic Atomizer Effects on Values of EC and pH of Nutrient Solution (Lakhiar et al., 2018b).

Lakhiar, Imran Ali et al., School of Agricultural Equipment Engineering, Jiangsu University, Zhenjiang Jiangsu, China.

How the main parameters of ultrasonic atomizers influence key properties of the atomized nutrient solution in an aeroponics system were studied. The Yamazaki tomato nutrient solution was selected as a nutrient example. In this test, spraying time and interval time were taken as quantitative factors with 12 levels (10, 20, 30, 40, 50, 60, 70, 80, 90, 100, 110, and 120 min, respectively), and ultrasonic atomizer frequency was taken as qualitative factor with 3 conditions (28 kHz, 107 kHz, 1.7 MHz). Based on test data, two regression formulations used to predict the values of ΔEC, and ΔpH of atomized Yamazaki tomato nutrient solution was established and inspected. The spraying interval time of ultrasonic atomizers had no significant effect on EC and pH of the atomized Yamazaki tomato nutrient solution; the ultrasonic atomizer frequency was more effective than spraying time on the values of EC and pH. Therefore, the high-frequency (1.7 MHz) ultrasonic atomizer is not suitable for aeroponics cultivation when using the Yamazaki tomato nutrient solution as aeroponics nutrient solution.

Russia

Optimization of the automated colorimetric measurement system for pH of liquid (Katin et al., 2017).

Katin Oleg et al., Don State Technical University, Rostov-on-Don, Russia.

A control system for the acidity of an aquatic environment was developed, which is relevant in such branches of agriculture as hydroponics and aeroponics. A method for measuring the pH of a liquid using a potentiometric method was studied. In particular, the article analyzes the advantages of this method over the use of a universal indicator paper and a color sensor for pH determination. It describes the components, the conditions of their operation and storage and the basic principles of the measuring system operation. The main goal of the development and application of the measuring system described in the article is to achieve a high degree of autonomy and automation for aeroponic and hydroponic greenhouse complexes.

Australia

Role of Salicylic Acid in Phosphite-induced Protection Against Oomycetes; a *P. cinnamomi—Lupinus augustifolius* Model System (Groves et al., 2015).

Groves, Emma et al., Centre of Phytophthora Science and Management, Veterinary and Life Sciences Murdoch University, Australia.

Salicylic acid (SA) as an alternative, or supplementary, treatment to be used to protect plant species was investigated. With the use of aeroponics chambers, foliar application of phosphite, SA, and phosphite/SA to Lupinus augustifolius was assessed in relation to root tip damage, in planta phosphite and SA concentration and lesion development. Both phosphite and SA were measurable at the root tip within 24 h of application, and all treatments significantly reduced the lesion length at 7 days. However, while phosphite and SA application increased the in plants SA concentration, phosphite caused significantly more damage to the root tip by reducing root cap layers and length than the SA, or phosphite/SA application. This study supports the notion that phosphite-induced sensitivity may be SA-dependent, as both phosphite and SA were found to control *P. cinnamomi* and stimulate SA accumulation.

Phosphite is used to control Oomycetes in a wide range of horticultural and native plant species worldwide. However, phosphite can be phytotoxic, and some pathogens have exhibited a reduction in the effectiveness of phosphite due to prolonged use.

China
Effects of Different Cultivation Patterns on Nutrient and Safety Qualities of Vegetables: A Review (Yue et al., 2015).

Hu Yue et al., Development Center of Plant Germplasm Resources, College of Life and Environmental Sciences, Shanghai Normal University, China.

This is a review article on different vegetable cultivation patterns studied world-wide in detail and the effects of these patterns on nutrient and safety qualities of vegetables. With the improvement of living standards, consumers' demand on large amounts of vegetables with different varieties, high quality and safety standards has promoted the rapid development of the vegetable industry, especially for the vegetable cultivation industry. Protected cultivations of vegetables have had a great growth in recent years, which was changing the vegetable cultivation methods from a traditional soil planting to diversified cultivation patterns, including but not limited to the greenhouse, substrate, hydroponics, and aeroponics cultivation systems.

Germany
Dynamics of Concentrations and Nutrient Fluxes in the Xylem of *Ricinus communis*—diurnal Course, Impact of Nutrient Availability and Nutrient Uptake (Schurr et al., 2001).

Schurr, Ulrich et al., Institute of Bio- and Geosciences Jülich, Germany.

The diurnal courses of nutrient transport in the xylem and their response to external availability of nutrients was studied. In soil culture, maximal concentrations in all analyzed substances were observed during night-time. Over experimental periods of up to 20 days, concentrations of some ions increased, most by accumulation in the soil. Stringent nutrient conditions were established in a novel pressure chamber. An aeroponic nutrient delivery system inside allows the sampling of xylem sap from intact plants under full control of the nutrient conditions at the root. Analysis of xylem transport under these highly defined conditions established that: (1) diurnal variations in concentrations and fluxes in the xylem are dominated by plant-internal processes; (2) concentrations of nutrients in the xylem sap are highly but specifically correlated with each other; (3) nitrate uptake and nitrate flux to the shoot are largely uncoupled; and (4) in continuous light, diurnal variations of xylem sap concentrations vanish. Diurnal variations of xylem sap composition and use of the new technique to elucidate xylem-transport mechanisms are discussed.

Indonesia
Pengaruh Tingkat EC (Electrical Conductivity) terhadap Pertumbuhan Tanaman Sawi (*Brassica juncea* L.) pada Sistem Instalasi Aeroponik Vertikal (Pratiwi et al., 2015).

Pusdima Rahma Pratiwi et al., Universitas Islam Negeri (UIN) Sunan Gunung Djati Bandung, Indonesia.

The response of mustard growth on the nutrient EC level in a vertical aeroponic system was studied. This research was carried out in Bekasi Timur. The design used was Randomized Block Design with four treatments that A (EC level 1 mS/cm), B (EC level 1.5 mS/cm), C (EC level 2 mS/cm), and D (EC level 2.5 mS/cm) with four replications. The results showed that EC level had effect on plant height, root length, and wet weight.

Singapore

Growth Irradiance Effects on Productivity, Photosynthesis, Nitrate Accumulation, and Assimilation of Aeroponically Grown Brassica alboglabra (He et al., 2015).

He, Jie et al., Natural Sciences and Science Education Academic Group, National Institute of Education, Nanyang Technological University, Singapore.

The aeroponic growth of Brassica alboglabraplants with full nutrients under full sunlight with average midday photosynthetic photon flux density (PPFD) of 1,200 μmol/(m²s) was studied. Thirty days after transplanting, plants were, respectively, subjected to 10 days of average midday PPFD of 1,200 (control, L1), 600 (L2) and 300 μmol/(m²s) (L3). Productivity, photosynthetic carbon dioxide assimilation and stomatal conductance were significantly lower in low-light (L2 and L3) plants than in high-light (L1) plants. Low light plants had the highest nitrate accumulation in the petioles. Low light also had an inverse effect on total reduced N content. After different light treatments, all plants were re-exposed to another 10 days of full sunlight. Low-light plants demonstrated their ability to recover their photosynthetic rate, enhance productivity and reduce the nitrate concentration. These results have led to the recommendation of not harvesting this popular vegetable during or immediately after cloudy weather conditions.

USA

Contamination of Ammonium-based Nutrient Solutions by Nitrifying Organisms and the Conversion of Ammonium to Nitrate (Padgett et al., 1993).

Pamela E. Padgett et al., USDA Forest Service Pacific Southwest Research Station, Riverside, CA, USA.

The conversion of ammonium to nitrate and contamination by nitrifying organisms in ammonium-based nutrient solutions was studied. Maize (*Zea mays*) and pea (*Pisum sativum*) were grown under greenhouse conditions in aeroponic, hydroponic, and sand-culture systems containing 2 mM ammonium chloride as the sole nitrogen source and evaluated for the activity of contaminating nitrifying organisms. In all three culture systems, root colonization by nitrifying organisms was detected within 5 days, and nitrate was detected in the nutrient solution within 10 days after seedling transfer. In sand culture, solution nitrate concentration reached 0.35 mM by the end of the 17-day experiment. Nitrate was found in significant quantities in root and shoot tissues from seedlings grown in ammonium-based nutrient solutions in all of the solution culture systems. Maize seedlings grown in an ammonium-based hydroponic system contained nitrate concentrations at 40% of that found in plants grown in nitrate-based solution. The results have implications for the use of ammonium-based nutrient solutions to obtain plants suitable for research on induction of nitrate uptake and reduction or for research using solution culture to compare ammonium versus nitrate fertilization.

Canada
Control of Relative Growth Rate by Application of the Relative Addition Rate Technique to a Traditional Solution Culture System (Stadt et al., 1992).

K.J. Stadt et al., Department of Botany University of Alberta Edmonton, Canada.

Experiments to adapt the relative addition rate (RAR) technique to a conventional solution culture system was conducted. A background concentration requirement of $36\,\mu M$ nitrogen (N), with other nutrients supplied in proportion to N, was necessary to produce a constant relative growth rate (RGR) of *Triticum aestivum* L. (wheat) at a low RAR. Solution pH changes were reduced by increasing the percentage of ammonium ions in the nitrogen supply, but the plants exhibited dry weight reductions and symptoms of toxicity above 30% ammonium.

USA
Effect of Nutrient Medium pH on Symbiotic Nitrogen Fixation by *Rhizobium leguminosarum* and *Pisum sativum* (Evans et al., 1980).

L. S. Evans et al., Laboratory of Plant Morphogenesis, Biology Department Manhattan College, Bronx, New York, USA.

Experiments to measure the pH-sensitive steps in nodulation and symbiotic fixation by *Pisum sativum* and isolate RP-212-1 of *Rhizobium leguminosarum* were performed. An aeroponic system with rigorous pH control was used to obtain numerous effective nodules. After exposure to various pH levels, the following responses were measured: (1) legume root growth and development, (2) survival and growth rate of a single effective bacterial isolate, (3) degree of nodulation, (4) rate of nitrogen fixation, (5) plant biomass, and (6) nitrogen content of plants. Both bacterial growth and root development were adequate at all pH levels from 4.4 to 6.6, but efficient nodulation and nitrogen fixation did not occur at pH 4.8 and below. The processes required for symbiosis were about 10 times as sensitive to acidity as either bacterial growth or root growth alone. Nodulation was the most acid-sensitive step.

OLIVE

Italy
Salt Stress Modifies Apoplastic Barriers in Olive (*Olea europaea* L.): A Comparison Between a Salt-Tolerant and a Salt-Sensitive Cultivar (Rossi et al., 2015).

Rossi, Lorenzo et al., BioLabs—Institute of Life Sciences,
Scuola Superiore Sant'Anna, Pisa, Italy.

Frantoio and Leccino were studied and found that they differentially avoid salt intrusion into the roots and translocation to the shoot. They investigated the anatomical changes in olive roots under salt stress and conducted X-ray microanalysis of frozen-hydrated samples that gave information on roots ion selectivity. They found genotypes-depended apoplastic adjustments in roots.

Italy

Ionic Relations of Aeroponically Grown Olive Genotypes, During Salt Stress (Tattini, 1994).

Tattini, Massimiliano, Istituto sulla Propagazione delle Specie Legnose, Consiglio Nazionale delle Ricerche Scandicci, Florence, Italy.

Two olive (*Olea europaea* L.) genotypes, 'Frantoio' and 'Leccino', by exposing them to increasing concentrations of NaCl (0–30–60–120 mM) in an aeroponic cultivation system for 60 days were studied. Dry weights and sodium and potassium contents of apical and basal leaves, new and old wood, and roots were measured to determine Na uptake rate, Na translocation rate and K-Na selectivity ratio (SK, Na). 'Frantoio' showed a higher salt resistance than 'Leccino'. 'Frantoio' and 'Leccino' had a similar Na uptake rate, but largely differed for Na translocation to the shoot. Furthermore 'Frantoio' exhibited a higher K-Na selectivity than 'Leccino' at both whole plant level and above all at the level of shoot system. Resistance mechanism of 'Frantoio' is probably related to Na exclusion by roots and to the ability to maintain an appropriate K/Na ratio in actively growing tissues.

PEA

USA

Aeroponics Chambers for Evaluating Resistance to Aphanomyces Root Rot of Peas (*Pisum sativum*) (Rao et al., 1995).

Rao, A. et al., University of Wisconsin-Madison, Madison, WI, USA.

An aeroponics system for evaluating resistance of peas to *Aphanomyces euteiches* was studied. Five pea genotypes were tested for resistance to two isolates of *A. euteiches* at concentrations of 0–10,000 zoospores per milliliter. Pea genotypes reacted differently to isolates of *A. euteiches*. Genotype 86–2231 demonstrated the lowest amount of disease and Dark Skin Perfection the highest. Isolate 467 produced consistently higher disease scores than isolate FB1. Inoculum concentration and isolate×concentration effects were significant. This study provides evidence that an aeroponics system can be useful in evaluating the reaction of peas to *A. euteiches*.

USA

Correlation of Pectolytic Enzyme Activity with the Programmed Release of Cells from Root Caps of Pea (*Pisum sativum*) (Hawes et al., 1990).

Martha C. Hawes et al., Department of Soil, Water and Environmental Sciences, The University of Arizona, Tucson, Arizona, USA.

The separation of the cells in pea (*Pisum sativum*) were studied using an aeroponic system in which separated cells were retained on the root until they were washed off for counting. It was found that cell separation is a developmentally regulated, temperature-sensitive process that appears to be regulated independently of root growth. No cells were released from very young roots. When plants were grown aeroponically, cell numbers increased with increasing root length to a mean of 3,400 cells per root, at which point the release of new cells ceased. The process could be reset and synchronized by washing the root in water to remove shed cells. Cell separation from the root cap was correlated with pectolytic enzyme activity in root cap tissue. Because these cells that separate from the root cap ensheath the root as it grows and thus provide a cellular interface between the root surface and the soil, it was propose the cells be called "root border cells."

PEANUT

Israel
Response of Peanuts (*Arachis hypogaea* L.) grown in Saline Nutrient Solution to Potassium Nitrate (Silberbush et al., 1988).

> *Silberbush, M. et al., Ben-Gurion University of the Negev, J. Blaustein Institute for Desert Research, Sede-Boqer, Israel.*

The use of potassium nitrate (KNO_3) as a means to reduce salt damage to peanuts (*Arachis hypogaea* L.) in sprayhydroponics (aeroponics) with 50 mM NaCl and 1, 3, 5, 7, and 9 mM KNO_3 was tested. The addition of KNO_3 caused an increase in dry weight of the shoot, root length, chlorophyll content, nitrate reductase activity, and the reduction of salt injury symptoms in the leaves with an optimum at 5–7 mM KNO_3. None of these factors was correlated with proline content. Mean root diameter was abnormally thick in low KNO_3 concentrations and became thinner in higher KNO_3 concentrations, reaching the lowest value at about 6 mM KNO_3. Concentrations of Na^+ and Cl^- in the shoot, but not in the root, decreased with the increase in KNO_3 concentration. It is suggested that $K+$ and NO_3-inhibited Na^+ and Cl^- translocation from the roots to the shoots so that leaf metabolism was protected against salt damage.

PEPPER

China
Pyramid Shaped Hydroponic and Aeroponic Technology—a New Technology for Pepper Cultivation (Chen et al., 2017).

> *Yinhua Chen et al., Lishui Academy of Agricultural Sciences, Lishui, China.*

This is a review paper on a pyramid-shape aeroponic system including the base construction, the process of cultivation, and the daily management requirements for growing pepper.

Evaluation of Hydroponic Techniques on Growth and Productivity of Greenhouse Grown Bell Pepper and Strawberry (Albaho et al., 2008).

Kuwait

Albaho, M. et al., Desert Agriculture and Ecosystems Program,
Kuwait Institute for Scientific Research, Kuwait City, Kuwait.

Two hydroponic techniques, i.e., nutrient film technique and A-shaped aeroponics, were evaluated and a closed insulated pallet system based on continuous subirrigation system with fertilizers in reservoirs to ensure a reserve within the root zone and compared them to the conventional soil-based cultivation method (control) in Kuwait. The experiment was conducted in an acrylic covered greenhouse having an evaporative cooling system with ambient temperatures ranging from 15°C to 20°C at night and 24°C to 35°C during the day throughout the period from October 2005 to May 2006. Vegetative growth, flowering, and fruiting of bell pepper (Capsicum annuum L. cv. Yara) and strawberry (Fragaria versca L. cv. Americana Porter) were evaluated. Yields were lower in the closed systems than for the control. There were significant differences between amounts of water consumed in the soil-less techniques with consumption ranging from 42.9% to 62.9% of the control for peppers and 54.3%–79.1% for strawberry.

PETUNIA

USA

Stem Versus Foliar Uptake during Propagation of *Petunia* x *hybrida* Vegetative Cuttings (Santos et al., 2009).

Santos, Kathryn M. et al., Department of Environmental
Horticulture, University of Florida, Gainesville, FL, USA.

Gene expression patterns associated with suberization at the stem scar region using aeroponic technology were identified. Changes in the relative abundance of different transcripts suggested a potential involvement of the plant hormone abscisic acid (ABA) in the wound-healing processes.

POTATOES

Indonesia

Seed Potato Production Using Aeroponics System with Zone Cooling in Wet Tropical Lowlands (Sumarni et al., 2013).

Sumarni, E. et al., Universitas Jenderal Soedirman,
Department of Agricultural Engineering, Indonesia.

The methodology to determine the appropriate cooling zone temperature on seed potato production in wet tropical lowlands was developed. Planting potatoes in wet tropical lowland is one alternative to help farmers and protect the environment. Cultivation techniques used in this study were the aeroponics system with three cooling zones

(10°C, 15°C, and 20°C) and control (room temperature). The 'Granola' potato seeds were used in the experiment, and were derived from tissue culture. The results showed that the cooling zone could reduce the stress on the potato roots cultivated by aeroponics in wet tropical lowlands, although burning wilt also happened due to high temperatures in the plant canopy. Further studies are needed to control the temperature at the top of the plant to prevent the plants from burning wilt due to high temperature stress.

UK

Response of three potato cultivars grown in a novel aeroponics system for mini-tuber seed production (Mateus-Rodruguez et al., 2012).

Mateus-Rodruguez, J. et al., Crop Sciences, University of Reading, UK.

Mini-tuber production based on a novel, rustic and publicly available aeroponics system at the International Potato Center (CIP) was developed. The technology is proposed as an alternative to conventional systems of pre-basic seed (mini-tuber) production that use soil-based substrates requiring bromide for sterilization. Previous research has shown that the aeroponics technology is potentially efficient for specific potato cultivars. The overall objective of this study was to evaluate plant growth and mini-tuber production of three potato cultivars grown in an aeroponics system under greenhouse conditions at CIP's experimental station in La Molina, Lima (Peru). The study was conducted between August 2008 and April 2009. The results showed that the aeroponics system is a viable technological alternative for the potato mini-tuber production component within a potato tuber seed system.

UK

Potato minituber production at different plant densities using an aeroponic system (Abdullateef et al., 2012).

*Abdullateef, S. et al., School of Agriculture, Policy
and Development, University of UK.*

The effect of plant density (25, 35, and 50 plants/m^2) on minituber production in aeroponics was investigated. The aeroponic cultivation system consisted of 18 sections covered by Styrodur® sheets. Nutrient solution was sprayed by fog nozzles every 5 min for 20 s. Potato plantlets (*Solanum tuberosum* L. 'Marfona') obtained by in vitro cultivation were acclimatised in the greenhouse and then planted in small rockwool cubes which were fixed in holes in the Styrodur® sheets.

Brazil/USA

Evaluation of "UFV Aeroponic System" to Produce Basic Potato Seed Minitubers (da Silva et al., 2018).

*A collaborative study between Federal University of
Viçosa, Brazil (de Silva et al) and Department of Botany,
University of California Riverside (M McGiffen).*

Misting nozzle types and the coating on the bucket's inner wall on the yield of basic potato minitubers were evaluated. Dry weight of roots, stems and leaves besides minituber number and tuber fresh weight were evaluated.

India

Effect of Micro-Plants Hardening on Aeroponic Potato Seed Production (Muthuraj et al., 2016).

Muthuraj et al., ICAR, Central Tuber Crops Research Institute, Kerala, India.

A study was conducted at the four potato research institutes (Thiruvananthapuram), to evaluate micro-plants hardening on aeroponic potato seed production in Bangladesh and its impact on its economics.

India

Feasibility Studies for Two Consecutive Crops of Potato (*Solanum tuberosum* L.) under Aeroponic System in North-Western Plains of India (Singh et al., 2017).

Singh et al., Central Potato Research Station and the Punjab Technical University, Jalandhar, Punjab, India.

Feasibility studies were conducted to optimize the use of aeroponic technology for fast multiplication of basic seed stocks from in vitro plantlets. To make better use of expensive aeroponics facility, two production cycles of consecutive crops per year were tried. Experiment was conducted at ICAR-Central Potato Research Station, Jalandhar, Punjab, on three varieties and two hardening mediums before planting to aeroponic system for two consecutive crops. All the three varieties performed well during main season but Kufri Surya emerged as the best performing variety under aeroponics during second consecutive season.

Ethiopia

Determination of Nutrient Solutions for Potato (*Solanum tuberosum* L.) Seed Production under Aeroponics Production System (Lemma et al., 2017).

Tessema Lemma et al., Ethopian Institute of Agricultural Research, Holette Agricultural Research Center, Ethopia.

Aeroponic nutrient solutions for the production of high yield clean potato seed at the Holetta Agricultural Research Center were determined. Each crop has an optimum nutritional requirement. Even each potato cultivar may require a specific nutrient solution in an aeroponics unit. It was determined that treatment B represented the optimum nutrient concentration rate to use in an aeroponics minituber production system under.

India

Review: Methods of Pre-Basic Seed Potato Production with Special Reference to Aeroponics—A Review (Buckseth et al., 2016).

Buckseth, T et al., Central Potato Research Institutes, Shimla, Kufri, Thiruvananthapuram Kerala, India.

This paper reviewed methods for pre-basic seed potato production at the Central Potato Research Institutes, Shimla, Kufri, Thiruvananthapuram, Kerala, India. The aeroponics technology provided the good quality seed of potato and mitigated the problem of shortage of good quality seeds and strategies for rapidly multiply the seed tubers. The cost of growing a tuber using aeroponics is about one-quarter the cost of a conventionally—grown tuber.

Turkey

Mini Tuber Production in Potato Via Aeroponic System (Hussein et al., 2018).

Hussein A. et al, Siirt University, Siirt, Turkey.

Evaluation of an aeroponic system for tuber production was conducted. The aeroponic system, which is independent of climatic conditions, has the advantage of improving the vegetative growth, delaying tuber formation, prolonging the vegetative period, increasing the tuber yield per plant and total tuber yield while decreasing the tuber weight. Due to the problems experienced in potato seedling tuber production in recent years, it emerged as an alternative production system for Turkey.

Brazil

Electrical Conductivity of the Nutrient Solution and Plant Density in Aeroponic Production of Seed Potato Under Tropical Conditions (Calori et al., 2017).

Calori, Alex Humberto et al., Instituto Agronômico de Campinas, Brazil.

The influence of electrical conductivity of the nutrient solution and plant density on the seed potato minitubers production in aeroponics system was studied. The Agata and Asterix cultivars were produced in a greenhouse under tropical conditions. For both cultivars, the highest yield was observed for the 100 plants·sqm density.

Indonesia

G0 Seed Potential of The Aeroponics Potatoes Seed In The Lowlands With A Root Zone Cooling Into G1 In The Highlands (Sumarni et al., 2016).

Eni Sumarni et al, Syiah Kuala University, Indonesia.

Due to the challenges of growing potato seeds in a temperate zone, this study was conducted in 2016 to evaluate a modified root zone cooling method of an aeroponic system to produce high quality of tuber seeds in lowland. The First Generations (G_0) of var. Atlantic and var. Granola were used as plant materials, and randomized block design (RBD) with four replications was applied in this research. The comparison of seed weight between G_0 and G_1 was about 10 g and 54 g on average, respectively. Since the size and weight of such G_1 could be categorized as Large (L) in term of commercial seed market, it's implied that the lowland modified aerophonic system could be nominated as a prospective method for producing G_0 tuber seed in the future.

United Arab Emirates
Evaluating Potential Production of Mid-Late Maturing Minituber of Potato Cultivars and Promising Clones under Aeroponic System (Hassanpanah, 2014).

D. Hassanpanah, Islamic Azad University, UAE.

Potential production of mid-late maturing mini-tuber clones and cultivars using an aeroponic system during 2011–2013 in both laboratory and greenhouse of Ardabil Sabalan Behparvar Company were investigated. Five clones and cultivars (three promising clones 397009-3, 397082-10, and 397081-1, and two cultivars, Khavaran and Agria) were evaluated in completely randomized designs with three replications. Mean mini-tuber weight per square meter in aeroponic system was 6.16 g.

Colombia/Peru
Genotype by Environment Effects on Potato Mini-Tuber Seed Production in an Aeroponics System (Mateus-Rodriguez et al., 2014).

Julián F. Mateus-Rodriguez et al., Corporación Colombiana de Investigación Agropecuaria (CORPOICA) and International Potato Center (CIP), Lima, Peru.

Experiments were conducted to evaluate the environmental effect on plant development and mini-tuber production of a diverse group of potato genotypes grown under an aeroponic system was carried out in greenhouses located at La Molina (Lima) and Huancayo (Junín). Five contrasting environments were set-up and evaluated. A combined Analysis of Variance was performed for the variables "days to tuber set", "days to senescence" and "plant height". The Venturana variety (T_2) was the best performing genotype with a total average mini-tuber "weight" of 644 g/plant while the Chucmarina variety (T_1) performed best for the variable "number of mini-tubers" with an overall average of 60.2 mini-tubers per plant. It was recommended that the environment and management should ideally be tailored to the genotype as this will result in significant yield gains.

Canada
A Low Nutrient Solution Temperature and the Application of Stress Treatments Increase Potato Mini-tubers Production in an Aeroponic System (Oraby et al., 2015).

Oraby, Hesham et al., Institute of Nutrition and Functional Food, Laval University, Québec, Canada.

This study was conducted to test the impact of the nutrient solution temperature and the application of different stress treatments at tuberization on several growth variables for two potato cultivars grown in an aeroponic system. At 25 days after transplanting, low pH level, wilting stress and nitrogen withdrawal under warm (24°C) and cool (18°C) root-zone temperatures were applied to potato plants cv. Mystere and Chieftain. Significant differences in mini-tuber production, stolon number and length and root length were observed among root zone temperatures, stress treatments and cultivars. The results of this study demonstrate that a judicious use of these stresses can effectively promote tuberization in aeroponics.

Spain
Differential Growth Response and Minituber Production of Three Potato Cultivars under Aeroponics and Greenhouse Bed Culture (Tierno et al., 2014).

Tierno, Roberto et al., Basque Institute of Agricultural Research and Development, Spain.

Different methods and cultivars for the production of quality prebasic seed in potato were evaluated and compared. Two cultivation systems, aeroponics and greenhouse beds with a peat moss substrate, and three potato cultivars with different vegetative cycle, Agria, Monalisa and Zorba, were assayed. Plants in the aeroponic system showed increased growth and their vegetative cycle extended between 12% and 36% compared to the plants cultivated in greenhouse beds. Further studies are needed to optimize aeroponics system, which can be considered a high yield potato multiplication system, particularly for early or mid season potato cultivars that may produce best quality minitubers.

Colombia, Peru, Ecuador, Brazil
Technical and Economic Analysis of Aeroponics and Other Systems for Potato Mini-Tuber Production in Latin America (Mateus-Rodriguez et al., 2013).

J. R. Mateus-Rodriguez et al., Tibaitatá Research Center, Corporación Colombiana de Investigación Agropecuaria, Mosquera, Colombia, the International Potato Center (CIP), Avenida La Molina, Lima, Peru, International Potato Center, Quito, Ecuador, Empresa Brasileira de Pesquisa Agropecuaria, Clima Temperado, Pelotas, RS, Brazil, Instituto Nacional Autónomo de Investigaciones Agropecuarias, Quito, Ecuador.

This study systematized the technical and economic aspects of aeroponics and provided a benchmark comparison of this technology with other mini-tuber production systems as developed in Latin America: conventional, semi-hydroponics, and fiber-cement tiles technology. Research methodologies included: three-year registration of cash flows and production registers of aeroponics, economic and technical surveys, in-depth inquiry with managers of technologies. The results show that aeroponics as promoted by the International Potato Center (CIP) has several advantages, including high multiplication rates (up to 1:45), high production efficiency per area (>900 mini-tubers/m), savings in water, chemicals and/or energy, and positive economic indicators.

Indonesia
Effect of Electrical Conductivity (EC) in the Nutrition Solution on Aeroponic Potato Seed Production with Root Zone Cooling Application in Tropical Lowland (Sumarni et al., 2019).

Eni Sumarni et al., Departement of Agricultural Technology. Faculty of Agricultural. Jenderal Soedirman University, Indonesia.

Production of potato seeds aeroponically in wet tropical lowland with root zone cooling application has been successful. This study focuses on the impact of electrical conductivity (EC) concentration of potato seed growth and yield on aeroponics system with root zone cooling application in the lowland. The purpose of this research was to study and understand the effect of EC of the nutrition solution on growth and yield of potato seed on potato seed production aeroponically with root zone cooling application in tropical lowland. It was concluded that the best EC for potato seed production with aeroponics system root zone cooling application was EC 1 mS/cm in the first 3 weeks, and 4 mS/cm in the 4th week to harvest.

Indonesia

Aplikasi Zone Cooling pada Sistem Aeroponik Kentang di Daratan Medium Tropika Basah (Nurwahkyuningsih, 2013).

> *Nurwahkyuningsih et al., Program Studi Teknik Pertanian, Unsoed, Jl.dr. Soeparno Karangwangkal Purwokerto, 53113. Email: arny0879@yahoo.com.*

A study was conducted to obtain an appropriate cooling temperature on seed potato production in medium land. Aeroponic cultivation techniques used with three zone cooling (15°C, 19°C and 24°C) and controls. Potato varieties used in this study is Granola is from tissue culture. The results showed that the highest plant cooling obtained at day and night regions 19°C and 24°C at night. The highest number of leaves was obtained at 24°C day and night. The highest number of tubers obtained at 19°C day and night.

Malawi

Performance of different potato genotypes under aeroponics system (Chipanthenga et al., 2013).

> *Chipanthenga, Margaret et al., Ministry of Agriculture, Irrigation and Water Development, Department of Agricultural Research Services, Lilongwe, Malawi.*

This study was aimed at exploring the use of tissue culture and aeroponics techniques in the production of quality potato seed. Potato plantlets were produced in the tissue culture laboratory at Bvumbwe Research Station, Thyolo district of Malawi and then transferred to an aeroponics facility at Njuli Estate, Chiradzulu district of Malawi. Potato yields in developing countries are below potential yield because potato production is mainly constrained by lack of quality seed. Lack of potato seed systems to provide farmers with quality clean and certified potato seed has led majority of farmers save their own seed. Such potato seed is characterized by systemic viral and bacterial diseases that are transmitted from generation to generation and this leads to low crop yields. In this study, on an average 30 tubers were produced per plant under aeroponics system which is six times more than the conventional (use of soil-based substrate) seed potato production system under screen house conditions (5 tubers per plant) under Malawian conditions.

Spain

Comparison of hydroponic and aeroponic cultivation systems for the production of potato minitubers (Ritter et al., 2001).

Ritter, E. et al., NEIKER-Instituto Vasco de Investigación y Desarrollo AgrarioVitoria-Gasteiz, Spain.

Two cultivation systems, aeroponics and hydroponics in greenhouse beds for the production of potato minitubers were studied. Plants in the aeroponic system showed increased vegetative growth, delayed tuber formation and an extended vegetative cycle of about 7 months after transplanting. Therefore in 1999, two production cycles were obtained with the hydroponic system, but only one with the aeroponic system. However, compared with total production in hydroponics, the tuber yield per plant in the aeroponic system was almost 70% higher and tuber number more than 2.5 fold higher.

Spain

Potato minituber production using aeroponics: Effect of plant density and harvesting intervals (Farran et al., 2006).

Farran, Imma et al., Department of Plant ProductionInstitute of Agrobiotechnology, Navarra, Spain.

A study was conducted to optimize minituber production through aeroponics. Potato plantlets, cv Zorba, were grown aeroponically at two different plant densities (60 and 100 plants/m^2). Plants showed an extended vegetative cycle of about 5 months after planting. A higher number of stolons was obtained at low plant densities. The best productivity obtained in this study was 800 minitubers/m^2 for weekly harvests and a low plant density (60 plants/m^2).

China

Manipulating Aeroponically Grown Potatoes with Gibberellins and Calcium Nitrate (Wang et al., 2018).

Wang, Cui-Cun et al., Sichuan Academy of Agricultural Sciences Crop Research Institute, Chengdu, China.

A study was conducted to maximize tuber number and yield in an aeroponics production system. Manipulating the plants hormonal makeup is a strategy. GA3(GA), Calcium Nitrate (CaN) and their combination were foliarly applied several times to potato plants of cvs. Favorita and Mira during both the autumn and spring growing seasons in a semi tropical setting in Sichuan. For the cv. Favorita, CaN+GA and GA increased tuber weight per plant by 63% and 49% in autumn and 53% and 73% in spring, respectively, as compared to the control(CK); tuber number per plant under CaN+GA and GA treatments increased by 90% and 85% in autumn and 35% and 52% in spring, respectively. Tuber numbers over

1,400/m² were harvested with cv. Mira during the spring season with the GA and CaN+GA treatments.

Indonesia

An Early Detection of Latent Infection of *Ralstonia solanacearum* on Potato Tubers (Baharuddin, 2014).

Baharuddin et al., Department of Plantation Crop, Agriculture Polytechnic Pangkep, Pangkep, Indonesia.

A study was conducted to determine an early and accurate detection of Ralstonia solanacearum in potato seedsource production (G0) aeroponics cultivation system, which previously was treated with microbial antagonists and artificially infected with Ralstonia solanacearum. Ralstonia solanacearum is not only a soil-borne pathogen but also a seedborne bacterial pathogen. Although the symptoms of wilt disease appeared in plants, "the seeds did not show typical symptoms. The result of polymerace chain reaction (PCR) using OLI1 and Y2 primers on tubers without antagonist with 100% disease incidence showed, that all of the tubers have positive results. From visually healthy seeds, only a few of them have positive result and mostly other tubers showed negative result. This means that the role of antagonists on suppression of wilt disease, cannot totally guarantee the tubers free from latent infection of R. solanacearum.

China

Study on Optimization of Hydroponic Technology of Virus-free Potato Plantlets in Winter in Chengdu Plain (Sang et al., 2014).

Sang, Youshun; et al., Chengdu Academy of Agricultural and Forestry Services, Chengdu, China.

A study was conducted on the effects of different heating measures on breeding virus-free potato plantlets with hydroponic technology. The virus-free plantlets of potato (Favorita) were used as test material. The potato seedlings grown in heaters and thermal insulation device-installed seedbed rooted 6.4 d earlier and emerged 13 d earlier compared to that in traditional seedbed. The growth of potato seedlings grown in heaters and thermal insulation device-installed seedbed was more robust. Their root length, stem diameter, plant height, leaf number and fresh weight were all higher than that of traditional potato seedlings. Moreover, the fruiting of aeroponics seedlings is also advanced. The heating hydroponics plays an important role in actual production.

South Korea

Nutritional and Structural Response of Potato Plants to Reduced Nitrogen Supply in Nutrient Solution (Chang et al., 2016).

Chang, Dong et al., Highland Agriculture Research Institute National Institute of Crop Science, South Korea.

In 2008 and 2014, the nutritional and structural response of potato leaves to reduced nitrogen (N) supply in the nutrient solution was investigated. Tissue culture plantlets of cultivars Atlantic and Superior were transplanted into a recirculating aeroponic system and grown using various nitrogen concentrations and at different time sequences compared with control plants which were grown under a constant nitrogen concentration. Potato shoots grown under reduced nitrogen supply exhibited suppression of total nitrogen (N), calcium (Ca) and magnesium (Mg) uptake, and enhancement of phosphorus (P) uptake in the leave tissues. The suppression of N uptake decreased shoot growth and leaf mesophyll development with inhibited chlorophyll accumulation. The data indicated that careful control of nitrogen concentration is necessary to minimize possible decrease in tuberization and tuber growth, especially for the cultivar Atlantic and during the spring season.

South Korea
Growth and Tuberization of Hydroponically Grown Potatoes (Chang et al., 2012).

Chang, Dong et al., Highland Agriculture Research Institute
National Institute of Crop Science, South Korea.

Three hydroponic systems (aeroponics, aerohydroponics, and deep-water culture) were compared for the production of potato (*Solanum tuberosum*) seed tubers. Aerohydroponics was designed to improve the root zone environment of aeroponics by maintaining root contact with nutrient solution in the lower part of the beds, while intermittently spraying roots in the upper part. Root vitality, shoot fresh and dry weight, and total leaf area were significantly higher when grown in the aeroponic system. This better plant growth in the aeroponic system was accompanied by rapid changes of solution pH and EC, and early tuberization. The first tuberization was observed in aeroponics on 26–30 and 43–53 days after transplanting for cvs Superior and Atlantic, respectively. Tuberization in aerohydroponics and deep-water culture system occurred about 3–4 and 6–8 days later, respectively. The number of tubers produced was greatest in the deep-water culture system, but the total tuber weight per plant was the least in this system. For cv. Atlantic, the number of tubers <30 g weight was higher in aerohydroponics than in aeroponics, whereas there was no difference in the number of tubers >30 g between aerohydroponics and aeroponics.

Brazil
Potato Basic Minitubers Production in Three Hydroponic Systems (Factor et al., 2007).

Factor, Thiago et al., Universidade Estadual Paulista, Brazil.

The production of potato basic minitubers in three hydroponic systems: aeroponic, DFT (deep flow technique) and NFT (nutrient film technique), using cultivars Agata and Monalisa were evaluated. The experiment was carried out under protected cultivation, from May to September, 2005, in Brazil, in a completely randomized block design, with split-plots. Tuber fresh weight and longitudinal diameter were not influenced by the hydroponic systems, with average values of 6.2 g and 30.2 mm, 6.3 g and 30.0 mm, 6.8 g and 31.0 mm, for hydroponic systems NFT, DFT and aeroponic, respectively. Minituber yield per plant and m² was significantly higher in the aeroponic system (49.3 e 874.4) in comparison the NFT (39.5 e 246.6) and DFT systems (41.6 e 458.0), respectively.

China

Effects of Elevated CO Applied to Potato Roots on the Anatomy and Ultrastructure of Leaves (Sun et al., 2011).

> *Sun, Z. et al., Institute of Plant Protection, Shandong Academy*
> *of Agricultural Sciences and General Station for Plant*
> *Protection of Shandong Province, Jinan, China.*

The root system of potato plants were treated with different oxygen and carbon monoxide concentrations for 35 days in aeroponic culture. Using 5% and 2% oxygen in the root zones, the thickness of leaves and palisade parenchyma significantly increased at 3,600 μmol(CO) mol in the root zone, compared with CO concentration 380 μmol mol or low CO concentration (100 μmol mol). In addition, smaller cells of palisade tissue, more intercellular air spaces and partially two layers of palisade cells were observed in the leaves with root-zone CO enrichment. Furthermore, there was a significant increase in the size of chloroplasts and starch grains, and the number of starch grains per chloroplast due to elevated CO only under 21% oxygen. The accumulation of starch grains in the chloroplast under elevated CO concentration could change the arrangement of grana thylakoids and consequently inhibited the absorption of sun radiation and photosynthesis of potato plants.

South Korea

Growth and Yield Response of Three Aeroponically Grown Potato Cultivars (*Solanum tuberosum* L.) to Different Electrical Conductivities of Nutrient Solution (Chang et al., 2011).

> *Chang, Dong et al., Highland Agriculture Research Institute*
> *National Institute of Crop Science, South Korea.*

Appropriate electrical conductivity (EC) recommendations were developed for the nutrient solutions used in hydroponic seed potato production for two new potato (*Solanum tuberosum* L.) cultivars with different maturing characteristics, Haryeong and Jayoung. The two cultivars, along with the standard cultivar, Superior, were grown with four nutrient solutions differing in EC (0.6, 1.2, 1.8, and 2.4 dS m). The EC of nutrient solution did not influence photosynthetic rate (Pr) or water use efficiency (WUE) of leaves, whereas it affected total chlorophyll content, transpiration rate (Tr), and stomatal conductance (Sc) in some cultivars, highest at 0.6 dS m EC for Tr and Sc, and lowest at the same EC for total chlorophyll content. The nutrient content of shoots was significantly affected by ECs of the nutrient solution. The total-N content was higher in 2.4 dS m EC, whereas a lower P, Ca, and Mg content was measured in plants grown at the same EC. Plants grown with a lower EC had inhibited shoot and stolon growth and earlier tuber formation compared to those grown with higher ECs. For higher EC treatments, tubers were observed 5–6 days later, particularly for the new cultivars. At harvest, the number of tubers produced from Superior and Jayoung plants was not affected by EC, whereas those of Haryeong increased at 1.8 dS m EC. The results suggest that potato cultivars Superior and Jayoung are able to grow at a wide range of solution EC levels, but that the new cultivar Haryeong needed optimal management of solution EC at 1.8 dS m to yield higher tuber production under an aeroponic cultivation system.

India

Effect of Supporting Medium on Photoautotrophic Microplant Survival and Growth of Potato (*Solanum tuberosum*) (Kaur et al., 2016).

*Kaur, RP et al., ICAR-Central Potato Research
Station, Jalandhar-(Punjab), India.*

The effects of different supporting medium on photoautotrophic micropropagation PAM in potato (*Solanum tuberosum* L.) using higher carbon dioxide concentration and ample light intensity were studied. Four different supporting mediums were evaluated including peat moss (PM), sand (SD), perlite (PL), vermiculite (VM) and liquid nutrition medium (NM) containing no supporting matrix as control. There were significant differences between the different supporting mediums used. Peat moss not only produced higher survival percentage (67.7%), but also significantly higher shoot length, root length, leaf area and fresh weight. Control (NM) was observed to be mostly at par to the other treatments sand (SD) and vermiculite (VM). Perlite was significantly inferior producing lesser shot lengths and fresh weight. Therefore, peat moss may be used as a suitable supporting medium for PAM in potato. The method reported in the study is simple, easy, effective and economical for micropropagation of potato and may be used for supplementing the Hi-tech potato seed production involving tissue culture and aeroponics technologies.

South Korea

Physiological Growth Responses by Nutrient Interruption in Aeroponically Grown Potatoes (Chang et al., 2008).

*Chang, Dong et al., Highland Agriculture Research Institute
National Institute of Crop Science, South Korea.*

A study was conducted to retard potato (*Solanum tuberosum* L. cultivars Superior, Atlantic, and Jasim) shoot growth by nutrient interruption and thereby induce tuber formation in an aeroponic cultivation system. In the period between 25 and 55 days after transplanting (DAT), a 10-day nutrient interruption was carried out on the potato plants. The interruption of nutrient supply significantly increased root activity and stolon anthocyanins, but it decreased photosynthesis and transpiration rates, and nutrient uptake in leaves. These changes in physiological growth responses induced tuber formation along the short, thin, and purple-pigmented stolons. Nutrient interruptions decreased shoot growth by 50%–60% and thereby decreased tuber growth 5%–36%, when compared with the control plants.

Kenya

Multiplication of Seed Potatoes in a Conventional Potato Breeding Programme: A Case of Kenya's National Potato Programme (Muthoni et al., 2014).

Muthoni, Jane et al., Kenya Agricultural Research Institute (KARI), National Potato Research Centre, Tigoni, Kenya.

This is a review article on breeding programmes for good cultivars to reduce the cost and encourage innovation. In the Kenyan potato breeding programme, at least 10 tons of seed potato tubers are required before a new potato cultivar can be officially released. Potato tuber bulking methods that have a high multiplication rate and at the same time reduce the number of field plantings should be adopted in order to produce planting materials of high health standards. Use of aeroponics system appears to be a good method for producing both basic seeds from true potato seeds or bulking of existing cultivars prior to production of certified seeds.

Kenya

Alleviating Potato Seed Tuber Shortage in Developing Countries: Potential of True Potato Seeds (Muthoni et al., 2013).

Muthoni, Jane et al., Kenya Agricultural Research Institute (KARI), National Potato Research Centre, Tigoni, Kenya.

The production of potatoes in developing countries was studied. Potato production in developing countries is hampered by constraints such as low soil fertility, pest and diseases and inadequate supply of good quality seed tubers. Lack of good quality seed is mostly a consequent of the prevailing seeds system; in most developing countries, majority of farmers recycle their own seeds or get them from informal sources. This leads to seed degeneration and buildup of tuber-borne diseases and hence low yields. In mitigating the problem of shortage of good quality seeds, strategies to rapidly multiply the seed tubers such as tissue culture in conjunction with hydroponics and aeroponics have been tried. However, these are expensive for most developing countries and thus unsustainable. Use of true potato seeds (TPS) is a technology that might solve the problem once and for all. The seeds are cheap, easy to carry and store, can be stored for long and do not transmit most diseases. The technology needs to be given serious thought and should be promoted in most developing countries so as to increase potato yields.

USA

Solid Matrix and Liquid Culture Procedures for Growth of Potatoes (Tibbitts et al., 1994).

Tibbitts, T.W. et al, University of Wisconsin, Department of Horticulture, USA.

The advantages and limitations of several different procedures for growth of potatoes for controlled ecological life-support system (CELSS) in space were reported. Solution culture, in which roots and stolons are submerged, and aeroponic culture were not found useful for potatoes because stolons did not produce tubers unless a severe stress was applied to the plants. In detailed comparison studies, three selected culture systems were compared, nutrient film technique (NFT), NFT with shallow media, and pot culture with deep media. It was concluded that there are serious limitations to the use of NFT alone for growth of potatoes in a CELSS system. These limitations can be minimized by using a modified NFT with a shallow layer of media, such as arcillite, yet additional work is needed to ensure high tuber production with this system under long photoperiods.

India
Arbuscular Mycorrhizal Fungal Inoculum Production Using *Ipomoea batata* Hairy Roots in Bioreactor (Chandran et al., 2009).

Chandran, R. et al., Department of Biotechnology, K.V.M. College of Engineering and Information Technology, Kerala, India.

Arbuscular mycorrhizal fungi (AMF) and their role in plant nutrition, protection against plant pathogens and soil quality were studied. The multiplication of AMF is possible only through the conventional pot culture technique and aeroponic culture system. Contamination is the major disadvantage of these techniques. To overcome these difficulties and to produce axenic culture of AMF, co-cultivation of AMF was done with spores of Glomus microcarpum var. microcarpum and transgenic hairy roots of *Ipomoea batata*, which were initiated through the mediation of *Agrobacterium rhizogenes* ATCC 15834. Mycorrhized I. batata hairy roots were grown in a simple bioreactor constructed for this purpose.

RADISH

Israel
Differences in Responses of Various Radish Roots to Salinity (Waisel et al., 1987).

Y. Waisel et al., Department of Botany, the George S. Wise Faculty of Life Sciences, Tel Aviv University, Tel Aviv, Israel.

Extension growth of tap-roots, initiation of lateral roots and extension growth of first lateral roots with young radish plants (Raphanus sativus cv. Rex) in aeroponic growth cabinets were investigated. Growth was not affected by 50 mM NaCl. At higher NaCl concentrations the three growth processes were adversely affected, but not at the same magnitude: Initiation of new laterals was found to be the most salt-tolerant process, whereas growth of laterals was the most sensitive one. Thus, it seems that even within one root system, different root-types respond independently to environmental stresses. Such individual responses may alter the appearance of the root system and, thus, be of ecological significance.

REVIEW

USA
Research Review Paper: Aeroponics for the Culture of Organisms, Tissues, and Cells (Weathers et al., 1992).

Weathers, P.J. et al., Department of Biology and Biotechnology, Worcester Polytechnic Institute, Worcester, MA, USA.

This paper reviewed the characteristics of aeroponics. Contrast is made, where appropriate, with hydroponics and aero-hydroponics as applies to research and commercial applications of nutrient mist technology. Topics include whole plants, plant tissue cultures, cell and microbial cultures, and animal tissue cultures with regard to operational considerations (moisture, temperature, minerals, gaseous atmosphere) and design of apparati.

RICE

China
Distribution and Speciation of Cu in the Root Border Cells of Rice by STXM Combined with NEXAFS (Peng et al., 2016).

Peng, Cheng et al., Chengdu University of Traditional Chinese Medicine, Chengdu, China.

The effects of copper (Cu) upon the Root border cells (RBC), as well as its distribution and speciation within the RBCs of rice (Oryza sativa L.) under aeroponic culture were investigated. RBCs serve plants in their initial line of defense against stress from the presence of heavy metals in the soil. In this research, light microscopy and synchrotron-based scanning transmission X-ray microscopy (STXM) combined with near edge X-ray absorption fine structure spectroscopy (NEXAFS) with a nanoscale spatial resolution were used. The results indicated that with increasing exposure time and concentration, the attached RBCs were surrounded by a thick mucilage layer which changed in form from an ellipse into a strip in response to Cu ion stress. Copper was present as Cu(II), which accumulated not only in the cell wall but also in the cytoplasm. It was believed that this is the first time that STXM has been used in combination with NEXAFS to provide new insight into the distribution and speciation of metal elements in isolated plant cells.

Phillipines
IRRI's drought stress research in rice with emphasis on roots: accomplishments over the last 50 years (Henry, 2013).

Henry, Amelia, International Rice Research Institute, Crop Environmental Sciences Division, Metro Manila, Philippines.

The activities of the IRRI from 1960 to the present were reviewed. IRRI was founded in 1960, and large efforts for research on root growth in response to drought were ongoing by the mid-1970s, with an emphasis on deep root growth, formation of coarse nodal roots, and the root pulling force method. In the 1980s, aeroponic studies on root morphology and anatomy and line-source sprinkler field studies were commonly conducted. The use of crosses to better understand the genetics of root traits started in the 1980s. Further characterization of the genetics behind root traits was conducted in the 1990s, specifically the use of molecular markers to select for root trait QTLs. A shift toward rainfed lowland experiments in addition to upland conditions began in the 1990s, with increased recognition of the different types of drought stress environments and characterization of root water uptake. In the 2000s, drought breeding efforts moved from selection of root traits to direct selection for yield under drought. In the 2010s IRRI identified two major drought-yield QTLs to be related to root traits, and phenotyping for association mapping of genes related to root traits and functions was underway. After direct selection for yield during the past decade that is now approaching impact at the farm level, it was found that root traits are indeed involved in improved yield under drought.

China
Developmental Characteristics and Response to Iron Toxicity of Root Border Cells in Rice Seedlings (Xing et al., 2008).

Xing, Cheng-hua et al., Jinhua College of Profession and Technology Bioengineering Institute Jinhua, China.

Experiments were conducted to investigate the ferrous iron (Fe^{2+}) effects on root tips in rice plant using border cells in vitro. The border cells were pre-planted in aeroponic culture and detached from root tips. Most border cells have a long elliptical shape. The number and the viability of border cells in situ reached the maxima of 1,600 and 97.5%, respectively, at 20~25 mm root length. This mortality was more pronounced at the first 1~12 h exposure to 250 mg/L Fe^{2+} than at the last 12~36 h. After 36 h, the cell viability exposed to 250 mg/L Fe^{2+} decreased to nought, whereas it was 46.5% at 0 mg/L Fe^{2+}. The decreased viabilities of border cells indicated that Fe^{2+} dosage and treatment time would cause deadly effect on the border cells. The increased cell death could protect the root tips from toxic harm. Therefore, it may protect root from the damage caused by harmful iron toxicity.

Japan
Hydraulic conductivity of rice roots (Miyamoto et al., 2001).

Naoko Miyamoto et al., Faculty of Agriculture, Tokyo University of Agriculture and Technology, Fuchu, Tokyo, Japan.

The hydraulic conductivities of rice roots (root Lpr/m^2 of root surface area) were measured using a pressure chamber and a root pressure probe technique. Young plants of two rice (Oryza sativa L.) varieties (an upland variety, cv. Azucena and a lowland variety, cv. IR64) were grown for 31–40 days in 12 h days with 500 μmol/(m²s) PAR and day/night temperatures of 27°C and 22°C. Root Lpr was measured under conditions of steady-state and transient water flow. Different growth conditions (hydroponic and aeroponic culture) did not cause visible differences in root anatomy in either variety. Values of root Lpr obtained from hydraulic (hydrostatic) and osmotic water flow were of the order of 10–8 m/(s MPa) and were similar when using the different techniques. In comparison with other herbaceous species, rice roots tended to have a higher hydraulic resistance of the roots per unit root surface area. The data suggest that the low overall hydraulic conductivity of rice roots is caused by the existence of apoplastic barriers in the outer root parts (exodermis and sclerenchymatous (fibre) tissue) and by a strongly developed endodermis rather than by the existence of aerenchyma. According to the composite transport model of the root, the ability to adapt to higher transpirational demands from the shoot should be limited for rice because there were minimal changes in root Lpr depending on whether hydrostatic or osmotic forces were acting. It is concluded that this may be one of the reasons why rice suffers from water shortage in the shoot even in flooded fields.

ROOTS

USA

An Intermittent Aeroponics System Adaptable to Root Research (Peterson et al., 1991).

Peterson, L.A. et al., Horticulture Dept., University of Wisconsin-Madison, Madison, USA.

A functionally reliable intermittent aeroponics system for plant growth studies was designed and implemented. The system provides the benefits of easy access to the root system and control of various parameters of the root environment. Misting the plant roots with about a 1/2 strength Hoagland's nutrient solution from 3 to 7 s at 10 min intervals has been sufficient for normal growth of a number of plant species. Installation of the aeroponics system into an environmental control chamber extends the capacity of the system to measure shoot and root responses in relation to controlled environmental parameters.

Israel

Aeroponics: A Search for Understanding Roots (Eshel and Waisel, 1997).

Eshel, A. et al., Department of Plant SciencesTel Aviv UniversityTel Aviv, Israel.

This is a chapter in a book on root formation and described a study of plant shoots. This formation was thoroughly studied and amply reported in the literature. Certain aspects of the functional significance of these shapes with regards to their ability to sustain mechanical strains and to capture light, were analyzed in detail.

South Africa

The Use of Aeroponics to Investigate Antioxidant Activity in the Roots of *Xerophyta viscosa* (Kamies et al., 2010).

Kamies, Rizqah et al., Molecular and Cell Biology, University of Cape Town, South Africa.

An aeroponic plant growth system was designed and optimized to observe the root's response to desiccation without the restrictions of a soil medium, allowing easy access to roots. Successful culture of both *X. viscosa* and the control, *Zea mays*, was achieved and dehydration stress was implemented through reduction of nutrient solution spraying of the roots. After drying to the air dry state (achieved after 7 days for roots and 10 days for shoots), rehydration was achieved by resumption of root spraying. *X. viscosa* plants survived desiccation and recovered but Z. mays did not. The root tissues appear to retain antioxidant potential during drying, for use in recovery upon rehydration, as has been reported for leaf tissues of this and other resurrection plants.

South Africa

The Changes in Morphogenesis and Bioactivity of *Tetradenia riparia*, *Mondia whitei*, and *Cyanoptis speciosa* by an aeroponic system (Kumari et al., 2016).

*Kumari, Aloka et al., Research Centre for Plant Growth
and Development, School of Life Sciences, University of
KwaZulu-Natal, Pietermaritzburg, South Africa.*

Antibacterial, antiplasmodial and cytotoxicity properties of 3-month-old *Tetradenia riparia*, *Cyanoptis speciosa* and *Mondia whitei* maintained in an efficient aeroponic system in a greenhouse were evaluated. The species were selected based on the importance of their roots in traditional medicine and their current conservation status. An aeroponic system provides an opportunity for improving quantity, quality, consistency, and biomass production of roots of medicinal plants. The study revealed that the aeroponic system could be used to produce clean, high-quantities and excellent quality of roots of *T. riparia*, *M. whitei* and *C. speciosa* for basic pharmacological research. In addition, aeroponic-grown *T. riparia* and *M. whitei* plants facilitate commercial and rapid propagation for therapeutic potential and conservation purposes. It could also satisfy traditional medicine and pharmaceutical industry demands.

USA

Aeroponic System for Control of Root-Zone Atmosphere (Kratsch et al., 2006).

*Kratsch, Heidi A. et al., Interdepartmental Plant Physiology Program,
Department of Horticulture, Iowa State University, USA.*

An aeroponic gas-delivery system was designed that allows statistical treatment of individual plants as discrete experimental units when root-zone-targeted treatments are imposed. Roots of intact plants were held in closed 1-L canning jars, with one plant per jar and the stem emerging from a hole drilled in the jar lid. Extra-fine fog nozzles controlled by a programmable timer sprayed roots periodically with nutrient solution. Gases were mixed and delivered to jars at a prescribed rate by using a system of capillary tubes. During a study of oxyge effects on formation of root nodules, 100% survival of plants and profuse nodulation resulted after 4 weeks. The system is easy to construct with inexpensive materials, has many applications for rhizosphere biology, and is unique among systems used previously.

Aeroponics is a viable alternative to other soil-less culture systems for maintaining plants with a controlled root-zone atmosphere.

USA

An Aeroponic Culture System for the Study of Root Herbivory on *Arabidopsis thaliana* (Vaughan et al., 2011).

*Vaughan Martha M et al., Department of Biological
Sciences, Virginia Tech, Blacksburg, VA, USA.*

An aeroponic culture system was developed based on a calcined clay substrate that allows insect herbivores to feed on plant roots while providing easy recovery of the root tissue. The culture method was validated by a root-herbivore system developed for *Arabidopsis thaliana* and the herbivore Bradysia spp. (fungus gnat). Arabidopsis root mass obtained from aeroponically grown plants was comparable to that from other culture systems, and the plants were morphologically normal. Bradysia larvae caused considerable root damage resulting in reduced root biomass and water absorption. After feeding on the aeroponically grown root tissue, the larvae pupated and emerged as adults. Root damage of mature plants cultivated in aeroponic substrate was compared to that of Arabidopsis seedlings grown in potting mix. The culture method will allow simple profiling and in vivo functional analysis of root defenses such as chemical defense metabolites that are released in response to below ground insect attack.

Austria

Implementation and Application of a Root Growth Module in HYDRUS (Hartmann et al., 2018).

Hartmann, Anne et al., Department of Water, Atmosphere, Environment, Institute of Hydrobiology and Aquatic Ecosystem Management, BOKU— University of Natural Resources & Applied Life Sciences, Vienna, Austria.

A root growth module was adapted and implemented into the HYDRUS software packages to model root growth as a function of different environmental stresses. The model assumes that various environmental factors, as well as soil hydraulic properties, can influence root development under suboptimal conditions. The implementation of growth and stress functions in the HYDRUS software opens the opportunity to derive parameters of these functions from laboratory or field experimental data using inverse modeling. One of the most important environmental factors influencing root growth is soil temperature. The effects of temperature in the root growth module was the first part of the newly developed HYDRUS add-on to be validated by comparing modeling results with measured rooting depths in an aeroponic experimental system with bell pepper.

USA

Unleashing the Potential of the Root Hair Cell as a Single Plant Cell Type Model in Root Systems Biology (eQiao et al., 2013).

Zhenzhen eQiao et al., University of Oklahoma, USA.

Procedures were optimized for obtaining root hair cell samples. They cultured the plants using an ultrasound aeroponic system maximizing root hair cell density on the entire root systems and allowing the homogeneous treatment of the root system. Then they isolated the root hair cells in liquid nitrogen. Isolated root hair yields could be up to 800–1,000 mg of plant cells from 60 root systems. Using soybean as a model, the purity of the root hair was assessed by comparing the expression level of genes previously identified as soybean root hair specific between preparations of isolated root hair cells and stripped roots, roots devoid in root hairs.

USA

Secondary Metabolism of Hairy Root Cultures in Bioreactors (Yoojeong et al., 2001).

Yoojeong, Kim et al., Department of Chemical Engineering
Worcester Polytechnic Institute, USA.

Vitro cultures as an alternative to agricultural processes for producing valuable secondary metabolites. Most efforts that use differentiated cultures instead of cell suspension cultures have focused on transformed (hairy) roots. Bioreactors used to culture hairy roots can be roughly divided into three types: liquid-phase, gas-phase, or hybrid reactors that are a combination of both. The growth and productivity of hairy root cultures are reviewed with an emphasis on success-ful bioreactors and important culture considerations. Together with genetic engi-neering and process optimization, proper reactor design plays a key role in the development of successful large-scale production of secondary metabolites from plant cultures.

USA

Transformed Roots of *Artemisia annua* Exhibit an Unusual Pattern of Border Cell Release (Weathers et al., 2001).

Weathers, Pamela et al., Department of Biology and Biotechnology,
Worcester Polytechnic Institute, Worcester, USA.

Border cells from *Artemisia annua* from hairy roots grown in shake flasks, cul-ture plates, a bubble column reactor, and a nutrient mist (aeroponic) reactor were examined. When well-hydrated roots were subjected to shear, border cells were first released as an agglomerate and did not disperse for several hours. Staining with neutral red and fluorescein diacetate (FDA) showed that both agglomerates and dispersed cells were alive. It was determined that FDA is cleaved by pectin methylesterase (PME) and that PME may not be particularly active in the released agglomerates until the border cells disperse. These results suggest that our hairy root clone is deficient in border cell release perhaps resulting from the transforma-tion process.

USA

Grenzwurzeln? (Hays, 1993).

Hays, Sandy Miller, US Department of Agriculture ARS Belsville, Maryland, USA.

This is a report on the German scientist Richard Zobel's study of 'grenzwurzeln' or boundary roots in the 1970s while working on mutant tomatoes and his use of aero-ponics in the experiments. He studied how plant roots take up soil elements.

USA

Using an Aerosol Deposition Model to Increase Hairy Root Growth in a Mist Reactor (Towler et al., 2007).

*Towler, Melissa J. et al., Department of Biology and Biotechnology,
Worcester Polytechnic Institute, Worcester, USA.*

The growth of hairy roots were studied using gas-phase reactors including a mist reactor. These reactors have distinct advantages over liquid-phase reactors including the ability to manipulate the gas composition, to allow effective gas exchange in a densely growing biomass, and to affect secondary metabolite production. Mathematical modeling suggested that roots in a mist reactor are often too sparsely packed to capture mist particles efficiently and cannot, therefore, meet the nutrient demands required to maintain high growth rates. Indeed, growth rates of *Artemisia annua* hairy roots increased significantly when the initial packing density increased or when a higher sucrose concentration was used in the medium.

Italy

Effects of pH in the Root Environment on Leakage of Phenolic Compounds and Mineral Ions from Roots of *Rosa indica* major (Abolitz et al., 1995).

M. Abolitz et al., Department of Horticulture, University of Florence, Italy.

The effects of pH on leakage of phenolic substances and ions from roots of rose (*Rosa indica* major) cuttings propagated in a nutrientless aeroponic fogging system were examined by incubation of root segments or intact roots in solutions of biological MOPSO+BTP buffer with pH 6.2, 7.6 or in water, following exposure to the buffers. The content of phenolic substances was measured spectrophotometrically at an absorbance of 260nm, whereas the contents of K^+ and N^+ were measured by means of Inductively Coupled Plasma-Atomic Emission Spectroscopy (ICP-AES) and the content of K^+ was also measured by flame photometry. The direct effects of high pH on root function, in addition to the indirect effects on the availability of mineral ions, are discussed.

SAFFRON

Poland

The Growth of Saffron (*Crocus sativus* L.) in Aeroponicsand Hydroponics (Souret et al., 2000).

*Souret, Frédéric F. et al., Institute of Genetics and
Biotechnology, University of Warsaw, Poland.*

The development of saffron (*Crocus sativus* L.) plants and the production of commercial saffron and saffron constituents were compared in three culture systems: aeroponics, hydroponics, and soil. On a dry weight, but not fresh weight basis, corm growth was increased in aeroponics and hydroponics as compared with growth in soil. Root length in aeroponics and hydroponics was reduced as compared with root length in soil, but shoot development was not significantly affected. Flowering was poor in all three culture systems, probably due to the small-sized (2.6 cm) bulbs used in propagating the plants. The production of stigmas and the concentration of the main constituents of saffron in the stigmas was similar in all three culture systems, suggesting saffron bulbs grown aeroponically and hydroponically may be used as a practical and renewable source of pharmacological compounds extracted from saffron.

SEED

USA
Seed Production from Aeschynomene Genetic Resources Rescued and Regenerated Using Aeroponics (Morris, 2014).

J.B. Morris, Plant Genetic Resources Conservation Unit,
Geneticist Plants, USDA, Griffin, Ga, USA.

An aeroponic system was evaluated for rescuing and regenerating photoperiod- and freeze-sensitive Aeschynomene accessions (cover crops). One-month-old seedlings from 11 accessions were field-planted in Griffin in 2012 and 2013. Four mature vegetative stem cuttings per accession, with at least 3 true leaves, were removed from plants that produced flowers but did not have enough time or photoperiod exposure to produce mature seed. Each cutting was placed in a hydroponic cloning machine. After approximately 2 weeks, 4 cuttings with healthy root systems were transferred to an aeroponic system. Productive plants yielding mature seed (22–584 seeds) were regenerated from 100% of the accessions. The success rate for cuttings was 71%. Seed weight and number were influenced by production year. Seed weights and numbers ranged from 0.630 to 2.317 g and 110–584, respectively, in 2012. In 2013, only the Mexican accession PI 544176 produced significantly higher seed weight (0.390 g) with the most seeds (203), compared to other accessions. However, PI 544176 production was lower compared to 2012 due to exposure to two hard freezes in the field that damaged plants enough to make them less vigorous in the aeroponic system. The aeroponic system's ability to produce productive plants demonstrated its usefulness for rescuing and regenerating seed from photoperiod- and freeze-sensitive Aeschynomene accessions.

SHALLOT

Indonesia
Aplikasi Root Zone Cooling System Untuk Perbaikan Pembentukan Umbi Bawang Merah (Allium cepa var. aggregatum) (Nurwahyuningsih, 2017).

Nurwahyuningsih et al., Departemen Teknik Mesin dan
Biosistem Institut Pertanian Bogor, Indonesia.

This study analyzed the effect of different root zone temperature to some extent the temperature is 10°C, 15°C, control and vernalization of plant growth and the formation of shallot bulbs by using aeroponic system. The experimental design used was a draft Plots Divided (Split Plot Design), which was arranged in a randomized block design with four replications. The main plot was a vernalization treatment (without vernalization and with vernalization). The subplots in the form of a nutrient solution temperature at 10°C, 15°C, and without cooling system as a control. The parameters measured were the number of leaves, the number of tillers, the number of bulbs, the weight of bulbs and the wet weight of root. There are no interaction between the annealing temperature by vernalization to the number of leaves, the bulb number, the weight of bulbs, and the weight of the roots. Cooling temperatures nutrient solution to improving root growth and bulb formation of shallot. Optimal root growth can improve nutrient uptaken by plants then can improve plant growth and bulb yield larger and heavier. Temperatures suitable for shallot cultivation in lowland tropical for producing tubers with quenching temperature is 10°C, nonvernalization.

SOCIAL IMPACT

USA

Aeroponic Gardens and Their Magic: Plants/Persons/Ethics in Suspension (Battaglia, 2017).

Battaglia, Debbora, Mount Holyoke College, South Hadley, MA, USA.

This paper delves into the question are plants 'for' persons or persons 'for' plants? Is it ethical to separate growing plants from earth/Earth and from earthlings? Where might 'responsible innovation' and 'innovative eclecticism' find a place in postgenomic discourse? And might a commitment to a dividual ethics guide lives in co-becoming to devise (as Latour recognizes in like terms) a scitech-diplomacy capable of resisting the programmatic pressures of new climatic regimes? She describes that out of the fire bombed ruins and food deprivation of the Second World War came one innovator's prototype for growing edible plants, suspended above earth and requiring a minimum of water. His aeroponic apparatus would later be referred to as The Genesis Machine, from the movie Star Trek II. This paper travels with roots in air into different spheres where the relations of persons, plants, and technology conspire to place future—growing into ethical suspension. The aim is to open questions for an anthropocenic future.

SOYBEANS

France

Charge Balance in NO_3^--fed Soybean. Estimation of K^+ and Carboxylate Recirculation (Touraine et al., 1988).

*Touraine, Bruno et al at the Biochimie et Physiologie Végétales,
Institut National de la Recherche Agronomique, Ecole
Nationale Supérieure Agronomique, Cedex, France.*

Soybeans (*Glycine max* L. Merr., cv Kingsoy) were grown on media containing NO_3^- or urea. The enrichments of shoots in K^+, NO_3^-, and total reduced N (N_r), relative to that in Ca^{2+}, were compared to the ratios K^+/Ca^{2+}, NO_3^-/Ca^{2+}, and N_r/Ca^{2+} in the xylem saps, to estimate the cycling of K^+, and N_r. The net production of carboxylates (R^-) was estimated from the difference between the sums of the main cations and inorganic anions. The estimate for shoots was compared to the theoretical production of R^- associated with NO_3^- assimilation in these organs, and the difference was attributed to export of R^- to roots. The net exchange rates of H^+ and OH^- between the medium and roots were monitored. The shoots were the site of more than 90% of total NO_3^- reduction, and N_r was cycling through the plants at a high rate. Alkalinization of the medium by NO_3^--fed plants was interrupted by stem girdling, and not restored by glucose addition to the medium. It was concluded that the majority of the base excreted in NO_3^- medium originated from R^- produced in the shoots, and transported to the roots together with K^+. As expected, cycling of K^+ and reduced N was favoured by NO_3^- nutrition as compared to urea nutrition.

USA
Detoxification and Evaluation of Foam Supports for Aeroponically Grown Soybean (Wagner et al., 1991).

Wagner, R. E. et al., Department of Agronomy,
University of Illinois, Urbana, IL, USA.

Using polyurethane-based foam plugs as supports this study aeroponically grew soybean plants [Glycine mu (L.) Merr.] that contain a phytotoxic compound. Plugs were detoxified after soaking in 1.0M NaOH for 1h and then in 950 mL/L ethanol for 12h. Five additional foam materials were evaluated as substitutes for polyurethane foam plugs. Four of the five foams tested (reticulated ether-based foam, etbafoam, Dowstrand foam, and L/C 200 foam) were nontoxic to soybean plants. Of these, reticulated ether-based foam (39.4 or 31.4 pores/cm) had physical properties similar to polyurethane foam and is recommended as a substitute for this material.

China
Immobilization of Aluminum with Mucilage Secreted by Root Cap and Root Border Cells is Related to Aluminum Resistance in *Glycine max* L (Cai et al., 2013).

Cai, Miaozhen et al., College of Geography and Environmental
Sciences, Zhejiang Normal University, Jinhua, China.

Two soybean cultivars differing in Al resistance were aeroponically cultured and the effects of Al on root mucilage secretion, root growth, contents of mucilage-bound Al for root tip Al and the capability of mucilage to bind Al were investigated. The root cap and root border cells (RBCs) of most plant species produced pectinaceous mucilage, which can bind metal cations. Increasing Al concentration and exposure time significantly enhanced mucilage excretion from both root caps and RBCs, decreased RBCs viability and relative root elongation except roots exposed to 400 µM Al for 48h in Al-resistant cultivar. Removal of root mucilage from root tips resulted in a more severe inhibition of root elongation. Of the total Al accumulated in root, mucilage accounted 48%–72% and 12%–27%, while root tip accounted 22%–52% and 73%–88% in Al-resistant and Al-sensitive cultivars, respectively. A 27Al nuclear magnetic resonance spectrum of the Al-adsorbed mucilage showed Al tightly bound to mucilage. Higher capacity to exclude Al in Al-resistant soybean cultivar is related to the immobilization and detoxification of Al by the mucilage secreted from root cap and RBCs.

USA

An Aeroponics System for Investigating Disease Development on Soybean Taproots Infected with *Phytophthora sojae* (Wagner et al., 1992).

Wagner, R. E. et al., Department of Plant Pathology, University of Illinois, Urbana-Champaign, Illinois, USA.

Pathogenesis on taproots of soybean (*Glycine max*) plants infected with *P. sojae* nondestructively with an aeroponics system was investigated. The main feature of the system is accessibility to the roots for direct observations, measurements, and application of inoculum. Lesion expansion on the taproots of the cultivar Corsoy was more rapid at 25°C than at 15°C. Lesion length after compatible interactions on the cultivars Corsoy, Sloan, and Agripro 26 was 8.0, 6.8, and 4.5 cm at 15°C and 13.5, 11.1, and 7.2 cm at 25°C, respectively, 9 days after inoculation. Cultivars were evaluated effectively for single-gene resistance based on the length and type of lesion formed 4 days after inoculation. Cultivars containing the genes Rps7 or Rps1 were susceptible, and cultivars containing the genes Rps1-b, Rps1-c, Rps1-d, Rps2, Rps3, or Rps6 were resistant to race 3 of the fungus.

USA

Use of Aeroponic Chambers and Grafting to Study Partial Resistance to *Fusarium solani* f. sp. glycines in Soybean (Mueller et al., 2002).

Mueller, D. S. et al., Crop Sciences, University of Illinois at Urbana-Champaign, Illinois, USA.

Plant introductions (PIs) and cultivars that are partially resistant (PR) to sudden death syndrome were studied. However, little is known about the nature of resistance to this disease. Seedlings of two PR PIs and two susceptible cultivars were inoculated with *Fusarium solani* f. sp. glycines in aeroponic chambers. Plants were inoculated by taping two sorghum seeds infested with F. solani f. sp. glycines to the main root. Foliar symptoms of the susceptible cultivars were higher than those on the PR PIs and were associated with lower root and plant dry weight. Root lesion lengths of the four soybean lines differed ($P < 0.05$), but did not correlate with foliar disease or any other variable. To better understand the resistance mechanism by distinguishing between root and plant resistance, three partially resistant PIs (PI 520.733, PI 567.374, and PI 567.650B) and one susceptible soybean cultivar (GL3302) were compared using different grafting combinations in aeroponic chambers. The results of sudden death syndrome evaluation indicated that resistance is conditioned by both the scion and the rootstock. All three PIs evaluated had resistance associated with the scion; resistance in PI 567.650B also was associated with the rootstock. Although the PR PIs used appear to have little or no root resistance, an aeroponic system and grafting may help identify new sources of resistance to F. solani f. sp. glycines with root- or whole-plant resistance.

SPACE APPLICATIONS

Poland
Concept of experimental platform to investigate aeroponic systems in microgravity conditions (Jurga et al., 2018a).

Jurga Anna et al., Wroclaw University of Science and Technology, Faculty of Environmental Engineering, Wroclaw, Poland.

The concept of experimental research system to investigate aeroponic cultivation in microgravity condition was studied. The main scientific objective is to define the forces acting in droplet-root system exposed to microgravity conditions especially the adhesion and cohesion phenomena. The concept of a research platform is presented in this paper and includes electrical, hydraulic and optical system.

USA
Planet Moon: The Future of Astronaut Activity and Settlement (Thangavelu, 2014).

Thangavelu, Madhu, Earth-Moon Cislunar Cycler, Department of Astronautical Engineering, School of Engineering, University of Southern California, Los Angeles, USA.

This paper explores the possibilities and extreme environmental challenges posed by human settlement of our closest planetary neighbour, the Moon.

Australia
Martian Base Agriculture: The Effect of Low Gravity on Water Flow, Nutrient Cycles, and Microbial Biomass Dynamics (Maggi et al., 2010a).

Maggi, Federico et al., School of Civil Engineering, The University of Sydney, Australia.

This paper a study of bioregenerative strategies for long-term life support in extraterrestrial outposts such as on Mars to determine whether soil-based cropping could be an effective approach for waste decomposition, carbon sequestration, oxygen production, and water biofiltration as compared to hydroponics and aeroponics cropping. However, it is still unknown if cropping using soil systems could be sustainable in a Martian greenhouse under a gravity of 0.38 g. The most challenging aspects are linked to the gravity-induced soil water flow; because water is crucial in driving nutrient and oxygen transport in both liquid and gaseous phases, a gravitational acceleration lower than $g = 9.806 \, m/s^2$ could lead to suffocation of microorganisms and roots, with concomitant emissions of toxic gases.

The effect of Martian gravity on soil processes was investigated using a highly mechanistic model previously tested for terrestrial crops that couples soil hydraulics and nutrient biogeochemistry. Net leaching of NO_3^- solute, gaseous fluxes of NH_3, CO_2, N_2O, NO, and N_2, depth concentrations of O_2, CO_2, and dissolved organic carbon (DOC), and pH in the root zone were calculated for a bioregenerative cropping unit under gravitational acceleration of Earth and for its homologous on Mars, but under 0.38 g. Martian cropping would require 90% less water for irrigation than on Earth, being therefore favourable for water recycling treatment; in addition, a substantially lower nutrient supply from external sources such as fertilizers would not compromise nutrient delivery to soil microorganisms, but would reduce the large N gas emissions observed in this study.

Australia

Space Agriculture in Micro- and Hypo-Gravity: A Comparative Study of Soil Hydraulics and Biogeochemistry in a Cropping Unit on Earth, Mars, the Moon, and the Space Station (Maggi et al., 2010b).

Maggi, Federico et al., School of Civil Engineering,
The University of Sydney, Australia.

A highly mechanistic model coupling soil hydraulics and nutrient biogeochemistry on soils on Earth (g = 9.806 m/s^2) was tested to highlight the effects of gravity on the functioning of cropping units on Mars (0.38 g), the Moon (0.16 g), and in the international space station (ISS, nearly 0 g). For each scenario, a comparison the net leaching of water, the leaching of NH_3, NH_4^+, NO_2^- and NO_3^- solutes, the emissions of NH_3, CO_2, N_2O, NO, and N_2 gases, the concentrations profiles of O_2, CO_2, and dissolved organic carbon (DOC) in soil, the pH, and the dynamics of various microbial functional groups within the root zone against the same control variables in the soil under terrestrial gravity was made.

USA

A Greenhouse for Mars and Beyond (Rahaim et al., 2008).

Rahaim, Christopher P. et al., HyperTech Concepts
LLC, Huntsville, Alabama, USA.

A detailed design study was conducted for a deployable greenhouse for Mars mission. The greenhouse has been designed so that it has a life span of at least 20 years, a leakage rate of no more that 1% of the total volume per day at the target working pressure of 50 kPa and provides at least six crew members with approximately 25% of their food supply. Artificial light is provided by high intensity red and blue light emitting diodes, but sunlight is also used by installing small Lexan windows on the rooftop. The greenhouse structure is a rigid graphite/epoxy sandwich structure with a footprint of 38 m^2. Radioisotope thermal electric generators are used to produce power for the greenhouse and its subsystems and the plants are grown in nested pockets located on vertical cylinders which allows for a growth area of 48 m^2. An aeroponic water and nutrient delivery system is used in order to reduce the greenhouse water usage. Harvesting and planting is achieved through the use of robotics specifically designed for this mission. The greenhouse structure and subsystems have a total weight of less than 10 metric tons. In this paper the design highlights of several of the subsystems of the greenhouse design were summarized.

Japan
A Spaceflight Experiment for the Study of Gravimorphogenesis and Hydrotropism in Cucumber Seedlings (Takahashi et al., 1999).

Takahashi, Hideyuki et al., Institute of Genetic Ecology, Tohoku University, Sendai, Japan.

The effect of modified gravity on the lateral positioning of a peg in cucumber seedlings in spaceflight was studied. It has been suggested that auxin plays an important role in the gravity-controlled positioning of a peg on the ground. Furthermore, cucumber seedlings grown in microgravity developed a number of the lateral roots that grew towards the water-containing substrate in the culture vessel, whereas on the ground they oriented perpendicular to the primary root growing down. The response of the lateral roots in microgravity was successfully mimicked by clinorotation of cucumber seedlings on the three dimensional clinostat. However, this bending response of the lateral roots was observed only in an aeroponic culture of the seedlings but not in solid medium. This system with cucumber seedlings is thus a useful model of spaceflight experiment for the study of the gravimorphogenesis, root hydrotropism and their interaction.

Japan
Trickle water and feeding system in plant culture and light-dark cycle effects on plant growth (Takano et al., 1987).

Takano, T. et al., Meijo University, Tempaku, Japan.

The use of rockwool, as an inert medium covered or bagged with polyethylene film for plant culture in space station was studied. The most important machine is the pump adjusting the dripping rate in the feeding system. Hydro-aeroponics may be adaptable to a space laboratory. The shortening of the light-dark cycles inhibits plant growth and induces an abnormal morphogenesis. A photoperiod of 12-hr-dark may be needed for plant growth.

SPRUCE

Sweden
Growth Responses of Rooted Cuttings from Five Clones of Picea abies (L.) Karst. after a Short Drought Period (Nordborg et al., 2001).

Nordborg, Fredrik et al., SLU, Southern Swedish Forest Research Centre, Alnarp, Sweden.

Rooted cuttings of five clones of Norway spruce [Picea abies (L.) Karst.] were studied in a well-watered and in an environment with a 5-day drought period. The study was performed an intermittent aeroponics system in a controlled environment chamber. During the root elongation was reduced to 1 mm/day compared with 3 mm/day in the treatment, but no significant reduction in shoot elongation was registered days after rewatering, the root elongation of drought-treated plants had recovered. At the end of the study, the increase in root length for the measured period did differ significantly between treatments, whereas the leading shoot length and biomass were in well-watered plants. The five clones, which had been selected for fast growth, similarly to the drought treatment and no interaction between clone and drought was In conclusion, drought affected the root growth directly, whereas the shoot growth was affected. This may reduce future growth as a result of a smaller leaf area and thereby a less assimilate production.

STRAWBERRY

Egypt

Effect of side and level of cultivation on production and quality of strawberry produced by aeroponic system (El-Behairy et al., 2003).

El-Behairy, U.A. et al., Ain Shams University, Cairo, Egypt.

Experiments were conducted at the Central Laboratory for Agricultural Climate Experimental Station, Dokki, Cairo, Egypt to investigate the effect of side and level of cultivation on production and quality of strawberry (Fragaria x ananassa) cv. Camarosa grown in aeroponic system. The experiment was carried out during two seasons (1999/2000 and 2000/2001). The experiment included two cultivation sides (north and south) of the aeroponic system and four cultivation levels (L1, L2, L3, and L4) in each side.

SUNFLOWER

Spain

Net simultaneous hydrogen and potassium ion flux kinetics in sterile aeroponic sunflower seedling roots: effects of potassium ion supply, valinomycin, and dicyclohexylcarbodiimide (Garrido, 1998).

Garrido, I. et al., U. E. Fisiologia, Vegetal, Facility Ciencias,
Universidad Extremadura, Badajoz, Spain.

This study involved the simultaneous monitoring of pH and potassium ion (K^+) concentration in a medium bathing 48 h sterile, aeroponic dark-grown sunflower (*Helianthus annuus* L.) seedling roots using specific noncombined high-sensitivity electrodes for pH and K(+). Net K(+) influx rates for different K(+) concentrations (0.25, 0.5, 1, 2, and 5 mM) lagged by approximately 60 min with respect to the, hydrogen ion (H^+) efflux, and showed the biphasic saturable kinetics (Epstein's Systems I and II) described by other authors.

Spain

Redox-related peroxidative responses evoked by methyl-jasmonate in axenically cultured aeroponic sunflower (*H. annuus* L.) seedling roots. (Garrido, 2003).

Garrido I, Fisiología Vegetal, Facultad de Ciencias,
Universidad de Extremadura, Badajoz, Spain.

The evocation of defense reactions was studied, as the oxidative burst in plants, substituting the elicitors or enhancing their effect. 48h dark- and sterilely cultured (axenic) aeroponic sunflower seedling roots excised and treated with different concentrations of methyl-jasmonate (MeJA) showed a strong and quick depression of the H(+) efflux rate, 1.80 μM MeJA totally stopping it for approximately 90 min and then reinitiating it again at a lower rate than controls. These results were wholly similar to those obtained with nonsterilely cultured roots and have been interpreted as mainly based on H(+) consumption for $O(2)(*-)$ dismutation to H_2O_2. Also, K(+) influx was strongly depressed by MeJA, even transitorily reverting to K(+) efflux. These results were consistent with those associated to the oxidative burst in plants.

Spain

Effect of some electron donors and acceptors on redox capacity and simultaneous net H^-/K^+ fluxes by aeroponic sunflower seedling roots: Evidence for a CN^--resistant redox chain accessible to nonpermeative redox compounds (Garrido, 1998).

Garrido, I., Fisiología Vegetal, Facultad de Ciencias, Universidad de Extremadura, Badajoz, Spain.

Studies were conducted on excised roots from aeroponic axenically 48h dark-grown sunflower (*H. annuus* L.) seedlings which showed redox activities. Being able to oxidize/reduce all the exogenously added electron donors/acceptors, that affected the H^+/K^+ net fluxes was simultaneously measured in the medium. Trials were performed with in vivo and CN^--poisoned roots; these showed null$^+/K^+$ net flux activity but still oxidized/reduced all the e^- donors/acceptors tested except NADH. NADH enhanced the rate of H^+ efflux by in vivo roots, otherwise not changing any of the normal flux kinetic characteristics, suggesting that NADH donates e^- and H^+ to the exocellular NADH oxidoreductase activity of a CN^--sensitive redox chain in the plasmalemma of the root cells.

TECHNOLOGY

Singapore

Aeroponics: Experiences from Singapore on a Green Technology for Urban Farming (Subramaniam et al., 2011).

R. Subramaniam et al., National Academy of Science and Nanyang Technological University, Singapore.

This is a chapter in an environmental leadership reference book on their experience with urban farming in Singapore.

Canada

Continuous Production of Greenhouse Crops Using Aeroponics (Nichols et al., 2002).

Nichols, M.A. et al. at the Department of Agriculture, Massey University, Canada.

An aeroponic production system that provides the opportunity to increase yield potential by up to 25%/year, without considering any potential advantages of the improved root environment is presented. The production system also provides a relatively simple technology for the easy management of stem numbers per unit area, so that the "plant density" can be adjusted to compensate for seasonal variations in solar radiation. Plants studied included tomato, cucumber, capsicum, egg plant, watermelon and honeydew and cantaloupe melon.

USA

Effect of Light Regimen on Yield and Flavonoid Content of Warehouse Grown Aeroponic *Eruca sativa* (Mattson et al., 2012).

Mattson, N.S. et al., Department of Horticulture, Cornell University, Ithaca, USA.

A study was conducted to determine whether LED arrays could provide necessary light quality and irradiance without affecting plant yield and quality. Flavonoids were the chosen qualitative proxy due to their known nutraceutical properties. A custom computerized LED array (Lighting Research Center, Rensselaer Polytechnic Institute, Troy, NY, USA) with an aeroponic growing system was used to provide differing intensity regimens of blue (460 nm, 8% of total light) and red (620 nm, 92% of total light) light to grow baby Arugula (*Eruca sativa* 'Astro'). Based on these results, LED arrays were found to be a suitable replacement to HPS, ideally with light levels increasing over developmental time.

China

Modern Plant Cultivation Technologies in Agriculture Under Controlled Environment: A Review on Aeroponics (Lakhiar et al., 2018a).

Imran Ali Lakhiar et al., Key Laboratory of Modern Agricultural Equipment and Technology, Ministry of Education, Institute of Agricultural Engineering, Jiangsu University, Zhenjiang, Jiangsu, People's Republic of China.

This is a review article on aeroponics. This review paper of the existing literature revealed as the population increases the demands for clean and fresh food increases alarmingly. People will turn to new plant growing technologies to fill up increasing food demands. Moreover, this review article concluded that aeroponics is a modern, innovative and informative technology for plant cultivation without incorporation of the soil. The system is the best plant growing technology in many aspects compared with different cultivation systems. The system is quickly increasing momentum, popularity and fastest growing sector of modern agriculture. It would be effectively used in various countries for vegetable production where natural resources are insufficient.

Indonesia

Applications of Temperature and Humidity Monitoring System at Aerophonic Plants Based on IoT (Saraswati et al., 2018).

Irma Saraswati et al., University of Sultan Ageng Tirtayasa, Indonesia.

A monitoring system was developed based on Android smartphone. This system was used to monitor temperature and humidity. Android-based monitoring system used java programming and was open source, besides the user can monitor the development of spinach plants, by looking at data that has been stored into the database so that users can track the development of the plant. This application can be accessed via smartphone with android operating system as a client server. Although the Indonesian agricultural system has improved significantly in recent years available agricultural land is becoming scarce. To overcome this problem and to reduce the impact of environmental climate changes one solution is the development of aeroponic system combined with IoT (internet of things).

Singapore
Farming of Vegetables in Space-limited Environments (He, 2015).

He, J, National Institute of Education, Nanyang Technological University, Singapore.

Farming systems were proposed for Singapore that not only increase productivity many-fold per unit of land but also produce all types of vegetables, all year-round for today and the future. This could be resolved through integrated vertical aeroponic farming system. These include low-cost urban community gardening and innovative rooftop and vertical farms integrated with various technologies such as hydroponics, aquaponics and aeroponics. This could involve manipulation of root-zone (RZ) environments such as cooling the RZ, modifying mineral nutrients and introducing elevated RZ carbon dioxide using aeroponics. It could also involve energy saving light emitting diodes (LEDs) for vertical aeroponic farming system to promote uniform growth and to improve the utilization of limited space via shortening the growth cycle, thus improving vegetable production in a cost-effective manner. Before 2011, vertical farms found in Japan, Korea, Holland, and England were all prototypes. Several commercial vertical farms have emerged in Singapore. The first vertical farm, Sky Greens was publicised in 2011, and commercialised in 2012. By growing vegetables in pots with soils under natural sunlight, Sky Greens mainly produces tropical vegetables. The greatest limiting factor for vegetable production in vertical farms is insufficient light for the plants.

Indonesia
Web-based Monitoring and Control System for Aeroponics Growing Chamber (Sani et al., 2016).

Sani, Muhammad Ikhsan et al., Computer Engineering Department, Faculty of Applied Science, Telkom University, Bandung, Indonesia.

A prototype system was designed and implemented for plant water and nutrients distribution to support the optimal application of an aeroponics system. It is based on a monitoring system which was used to observe the aeroponics growing chamber's parameters such as temperature, light, and pH. Meanwhile, the control system was used to manage actuators, i.e., mist maker and fan for delivering water moisture. Sensor data are transmitted via internet into a server in order to facilitate easier monitoring for users.

Thailand

Heat Pipe as a Cooling Mechanism in an Aeroponic System (Srihajong et al., 2006).

Srihajong, N. et al., Department of Mechanical Engineering,
Faculty of Engineering, Chiang Mai University, Thailand.

A mathematical model was developed explaining the operation of an aeroponic system for agricultural products. The purpose of this study was to determine the rate of energy consumption in a conventional aeroponic system and the feasibility of employing a heat pipe as an energy saver in such a system. A heat pipe can be theoretically used to remove heat from the liquid nutrient that flows through the growing chamber of an aeroponic system. To justify the heat pipe application as an energy saver, numerical computations have been done on typical days in the month of April from which maximum heating load occurs and an appropriate heat pipe set was theoretically designed. It was found from the simulation that the heat pipe can reduce the electric energy consumption of an evaporative cooling and a refrigeration system in a day by 17.19% and 10.34%, respectively.

China

Development of Droplets Penetrating Roots Performance Test Device and Tests Applied this Device in Ultrasonic Aeroponic System (Teng et al., 2014).

Yue Teng et al., Key Laboratory of Modern Agricultural Equipment and
Technology Ministry of Education and Jiangsu University, Zhenjiang, China.

Two ultrasonic nozzles were developed whose working frequencies were 1.7 MHz and 40 kHz, respectively, nozzle drive circuit, acquisition system of humidity and temperature. The objective was to be able to penetrate into the core of roots in an aeroponic system. This phenomenon is so-called "external is wet but internal is dry" and seriously affects aeroponic efficiency. It was concluded that for a cherry tomatoes root system density grown in an aeroponic system, droplets generated by 1.7 MHz ultrasonic atomizing nozzle were easier to penetrate into the core of the root; both size and concentration of droplets influenced droplets penetrating into the root.

India

Original Papers: IoT-based Hydroponics System Using Deep Neural Networks (Mehra et al., 2018).

Mehra, Manav et al., Department of Information Technology,
SRM Institute of Science and Technology, Chennai, India.

Control systems for aeroponic systems were developed by applying machine learning algorithms like Neural Networks and Bayesian network. Internet of Things allows for Machine to Machine interaction and controlling the system autonomously and intelligently. The system developed is intelligent enough in providing the appropriate control action for the hydroponic system based on the multiple input parameters gathered. A prototype for tomato plant growth as a case study was developed using Arduino, Raspberry Pi3 and Tensor Flow.

USA

Regeneration of Plants Using Nutrient Mist Culture (Weathers et al., 1988).

Pamela J. Weathers, et al., Department of Biology and Biotechnology, Worcester Polytechnic Institute, USA.

A nutrient mist used for in vitro culture of plant tissue in a novel bioreactor, wherein the tissues were grown on a biologically inert screen within a sterile chamber which allows excess media to drain away from the tissue was developed. Plants tested included Daucus, Lycopersicon, Ficus, Cinchona, and Brassica. The latter 4 genera were fully regenerated within the bioreactor. Cost analysis estimates showed a 65% savings in production costs (labor and materials) could be realized using nutrient mist culture. Nutrient mist culture offers significant improvements in the micropropagation of plants for aeroponic growing.

USA

An Intermittent Aeroponics System (Peterson et al., 1988).

Peterson, Lloyd A. et al., Horticulture Department College of Agricultural and Life Sciences, University of Wisconsin-Madison, USA.

An intermittent aeroponic system was developed and tested for potential use in plant growth studies. Cucumber (*Cucumis sativus* L.) were grown in the system from the cotyledon stage to flowering and fruiting within 40 days on a misting schedule of 7 s at 10 min intervals with about one half strength Hoagland's solution. The root environment was controlled at $20 \pm 1.1°C$. Root growth was rapid with good branching and root hair development. The proposed plant growth system provides the benefits of easy access to the root system and control of the misting solution composition as related to elemental concentration or to an application of a plant growth regulator at different stages in the developing root. It may also provide a system for testing short-term water stress in relation to plant development.

China

Prediction Model for Fundamental Frequency of Low Frequency Bending Vibration Piezoelectric Vibrator of Ultrasonic Atomizing Nozzle (Jianmin et al., 2015).

Gao Jianmin et al., College of Vehicles and Energy, Yanshan University, Qinhuangdao, China.

Ultrasonic atomization spraying technique was developed for small droplets, uniform size distribution, high roundness, large atomization quantity, and low liquid delivery pressure for application in aeroponics, agricultural humidifying fields, reagents atomization treatment, and semiconductor etching.

USA

The Vertical Farm: A Review of Developments and Implications for the Vertical City (Al-Kodmany, 2018).

Kheir Al-Kodmany, University of Illinois, Chicago, USA.

The need for vertical farms by examining issues related to food security, urban population growth, farmland shortages, "food miles", and associated greenhouse gas (GHG) emissions was presented. Urban planners and agricultural leaders have argued that cities will need to produce food internally to respond to demand by increasing population and to avoid paralyzing congestion, harmful pollution, and unaffordable food prices. The paper examines urban agriculture as a solution to these problems by merging food production and consumption in one place, with the vertical farm being suitable for urban areas where available land is limited and expensive. Luckily, recent advances in greenhouse technologies such as hydroponics, aeroponics, and aquaponics have provided a promising future to the vertical farm concept. Economic feasibility, codes, regulations, and a lack of expertise remain major obstacles in the path to implementing the vertical farm.

USA

One-Step Acclimatization of Plantlets Using a Mist Reactor (Correll et al., 2001).

Correll, M. J. et al., Department of Biology and Biotechnology,
Worcester Polytechnic Institute, USA.

A mist reactor was developed to grow and acclimatize carnation plants in vitro without using ex vitro acclimatization techniques. The acclimatization protocol in the reactor consisted of altering the mist-on period during the course of the culture period and a stepwise reduction in the relative humidity surrounding the plants from 98% to 70% relative humidity (RH) during the final week of in vitro growth. After transfer and further growth in an aeroponic greenhouse for 5 weeks, survival was 91% for plants grown in reactors, 81% from vented boxes, and 50% from unvented boxes. After 5 weeks in the greenhouse, the quantity of mid- and high-quality plants obtained from reactors and ventilated boxes was similar. Conditions in the mist reactor can be manipulated to produce plants that are readily acclimatized and are equal or better in quality and yield than plants produced using conventional methods.

USA

Method for Growing Plants Aeroponically (Zobel et al., 1975).

Zobel, R.W., Cabot Foundation, Harvard
University, Petersham, MA, USA.

A simple, inexpensive system was developed for growing plants with their roots bathed in nutrient mist. The aeroponics system uses a spinner from a home humidifier to propel nutrient solution into a polyethylene-lined plywood box atop which plants are supported on plastic light-fixture "egg crating." Success in growing a number of herbaceous and woody species, including nodulated legumes and nonlegumes, was reported.

China

Pyramid Shaped Hydroponic and Aeroponic Technology—a New Technology for Pepper Cultivation (Chen et al., 2017).

Yinhua Chen et al., Lishui Academy of Agricultural Science, China.

A pyramid shaped aeroponic system was developed to grow plants in an air or mist environment instead of using soils. The research reviewed pyramid-shaped hydroponic and aeroponic system and made conclusions from base construction, process of the cultivation and daily management.

Japan

Study of New Technology supporting design of Home Use Plant Factory (Li et al., 2014).

Li, Zhenpeng et al., Graduate School of Engineering, Chiba University, Japan.

A new technology was proposed for a plant factory for home use due to the growing demand of planting vegetables at home in China. Home plant factory presented a huge development potential and broad market prospect. This technology is an improvement of conventional aeroponics and it cannot only help grow vegetables easily and quickly, but also requires less water and energy than conventional hydroponics. In the research the UNPF (Ultrasonic Nebulizer Plant Factory) was compared with conventional hydroponics in the environment of indoor and to the application of the UNPF for the design of home plant factory. It was found that the potherb mustard and lettuce grew better in UNPF than in a control group.

TOMATO

China

Effects of Substrate-Aeration Cultivation Pattern on Tomato Growth (Zhao et al., 2010).

Zhao X et al., Liaoning Provincial Key Laboratory of Protected Horticulture, College of Horticulture, Shenyang Agricultural University, Shenyang, China.

A comparative study was conducted on the seedlings growth. Aeroponics can increase the fruit yield of tomato plant, but its cost is very high. In this paper, tomato seedlings were planted with three cultures, i.e., whole perlite culture (CK), perlite-aeration culture (T1), and aeroponics (T2). Compared with CK, T1 improved the gas environment in root zone significantly, with the carbon dioxide and oxygen concentrations in root zone being 0.2 and 1.17 times higher, and increased the plant height and stem diameter after 60 days of transplanting by 5.1% and 8.4%, respectively. The plant net photosynthetic rate of T1 was significantly higher than that of CK, with the maximum value after transplanting 45 days increased by 13%. T1 also increased the root activity and ion absorbing ability significantly, with the root activity after transplanting 45 days being 1.23 times of CK, and the root K, Ca, and Mg contents after transplanting 60 days increased by 31%, 37%, and 27%, respectively. The fruit yield of T1 was 1.16 times of CK. No significant differences in these indices were observed between T1 and T2, and less difference in the fruit soluble sugar and organic acid contents as well as the sugar-acid ratio was found among CK, T1, and

T2. It was suggested that perlite-aeration cultivation pattern was an easy and feasible way to markedly improve the fruit yield of tomato plant.

Israel
Roots of Tomato Respond to Mechanical Stimulation by Induction of Touch Genes. (Eshel et al., 2005).

Eshel, A. et al., Department of Plant Sciences, The George S. Wise Faculty of Life Sciences, Tel Aviv University, Tel Aviv, Israel.

A study was conducted on the molecular events associated with the response to mechanical impedance in plant shoots but not in their roots was studied. Plant roots growing in soil are subjected constantly to friction and mechanical resistance. The goal of the study was to identify components of the response mechanism to short and long term mechanical simulation of tomato root apices at the morphological and molecular level. Taking advantage of the unique aeroponic chambers of The Sarah Racine Root Research laboratory the effect of mechanical impedance on tomato roots was studied. The results show that mechanical stimulation brought about an increase in the root branching.

USA
Physiological and Molecular Responses of Aeroponically Grown Tomato Plants to Phosphorus Deficiency (Biddinger, 1998).

Biddinger et al., Department of Horticulture, Purdue University, West Lafayette, IN, USA.

The physiological changes occurring under phosphate (P) starvation to gene expression was studied. Phosphorus is one of the essential but limiting nutrients in nature. Roots of aeroponically grown tomato (*Lycopersicon esculentum* L.) plants were sprayed intermittently with nutrient solutions containing varying concentrations of P. Decreasing the concentration of P in the nutrient solution resulted in reduced biomass production and altered the tissue concentration of nutrients in roots and shoots. Phosphorus starvation increased the root:shoot biomass ratio and decreased net carbon dioxide assimilation and stomatal conductance. Phosphorus concentrations in roots and shoots decreased with decreasing concentration of P in the nutrient solution. P-deficient plants had a higher concentration of Ca in roots and Mg in shoots. Expression of the P starvation-induced gene, TPSI1, persisted even after 3 weeks of P starvation. The transcript accumulation in leaves was found to be a specific response to P starvation and not to the indirect effects of altered N, K, Fe, Mg, or Ca status. Accumulation of transcripts was also observed in stem and petioles, suggesting a global role for TPSI1 during P starvation response of tomatoes.

Spain
Net simultaneous hydrogen and potassium ion flux kinetics in sterile aeroponic sunflower seedling roots: effects of potassium ion supply, valinomycin, and dicyclohexylcarbodiimide (Garrido et al., 1998).

Garrido, Inmaculada et al., Plant Physiology, Extremadura University, Badajoz, Spain.

This study involved the simultaneous monitoring of variation of pH and potassium ion (K) concentration in a medium bathing 48 h sterile, aeroponic dark-grown sunflower (*H. annuus* L.) seedling roots using specific noncombined high-sensitivity electrodes for pH and K. Net K influx rates for different K concentrations (0.25, 0.5, 1, 2, and 5 mM) lagged by ≈60 min with respect to the, hydrogen ion (H) efflux, and showed the biphasic saturable kinetics (Epstein's Systems I and II) described by other authors.

USA
Localization of Ethylene Biosynthesis in Roots of Sunflower (*H. annuus*) Seedlings (Finlayson et al., 1996a).

Finlayson, Scott A. et al., Soil and Crop Sciences, College of Agriculture and Life Sciences, Texas A&M, Texas, USA.

Sunflower (*H. annuus* L.) seedlings were grown in aeroponic chambers which allowed for easy access to and easy harvesting of undamaged roots. In different portions of these roots we followed the rate of ethylene production, levels of 1-amin ocyclopropane-1-carboxylic acid (ACC), N-malonyl-ACC and ACC oxidase mRNA and activity of ACC oxidase. ACC oxidase was measured with an in vitro assay. ACC and N-malonyl-ACC by selected ion monitoring gas chromatography-mass spectrometry. Ethylene production was highest in the tip of the root and lower in the middle and basal (part nearest the hypocotyl) portions of the root. Treating the seedlings with ACC produced a rapid rise in ACC content and ethylene production and inhibited root elongation ACC oxidase activity was not induced by ACC treatment.

Canada
Drought-induced Increases in Abscisic Acid Levels in the Root Apex of Sunflower (Robertson et al., 1985).

J. Mason Robertson et al., University of Calgary, Calgary, Canada.

Abscisic acid (ABA) levels were measured in 3-mm apical root segments of slowly droughted sunflower plants (*H. annuus* L. cv Russian Giant) as the methyl ester by selected ion monitoring gas chromatography-mass spectrometry. An internal standard, hexadeuterated ABA (d6ABA) was used for quantitative analysis. Sunflower seedlings, grown in aeroponic chambers, were slowly droughted over a 7-day period. Drought stress increased ABA levels in the root tips at 24, 72, and 168 h sample times. Control plants had 57–106 ng/g ABA dry weight in the root tips (leaf water potential, −0.35 to −0.42 megapascals). The greatest increase in ABA, about 20-fold, was found after 72 h of drought (leaf water potential, −1.34 to −1.47 megapascals). Levels of ABA also increased (about 7- to 54-fold) in 3-mm apical root segments which were excised and then allowed to dessicate for 1 h at room temperature.

USA
The Effect of Carbon Dioxide on Ethylene Evolution and Elongation Rate in Roots of Sunflower (*H. annuus*) Seedlings (Finlayson et al., 1996b).

Finlayson, Scott A. et al., Soil and Crop Sciences, College of Agriculture and Life Sciences, Texas A&M, Texas, USA.

The effects of varying carbon dioxide (CO_2) concentrations were investigated on ethylene production by excised and intact sunflower roots (*H. annuus* L. cv. Dahlgren 131). Seedlings were germinated in an aeroponic system in which the roots hung freely in a chamber and were misted with nutrient solution. This allowed for treatment, manipulation and harvest of undamaged and minimally disturbed roots. While exposure of excised roots to 0.5% CO_2 could produce a small increase in ethylene production (compared to roots in ambient CO_2), CO_2 concentrations of 2% and above always inhibited ethylene evolution. This inhibition of ethylene production by CO_2 was attributed to a reduction in the availability of ACC; however, elevated CO_2 had no effect on 1-aminocylopropane-1-carboxylic acid (ACC) oxidase activity. ACC levels in excised roots were depressed by CO_2, at a concentration of 2% (as compared to ambient CO_2), but n-malonyl-ACC (MACC) levels were not affected.

TREES

Israel

Root–Shoot Allometry of Tropical Forest Trees Determined in a Large-Scale Aeroponic System (Eshel et al., 2013).

Eshel, Amram et al., Department of Molecular Biology and Ecology of Plants, Tel Aviv University, Tel Aviv, Israel.

The allometric relationships among the tropical tree organs, and carbon fluxes between the various tree parts and their environment were studied. Information on canopy-root interrelationships is needed to improve understanding of above- and below-ground processes and for modelling of the regional and global carbon cycle. Allometric relationships between the sizes of different plant parts will be determined.

Two tropical forest species were used in this study: Ceiba pentandra (kapok), a fast-growing tree native to South and Central America and to Western Africa, and Khaya anthotheca (African mahogany), a slower-growing tree native to Central and Eastern Africa. Growth and allometric parameters of 12-month-old saplings grown in a large-scale aeroponic system and in 50-L soil containers were compared. The main advantage of growing plants in aeroponics is that their root systems are fully accessible throughout the plant life, and can be fully recovered for harvesting. The expected differences in shoot and root size between the fast-growing C. pentandra and the slower-growing K. anthotheca were evident in both growth systems. Roots were recovered from the aeroponically grown saplings only, and their distribution among various diameter classes followed the patterns expected from the literature. Stem, branch and leaf allometric parameters were similar for saplings of each species grown in the two systems. The aeroponic tree growth system can be utilized for determining the basic allometric relationships between root and shoot components of these trees, and hence can be used to study carbon allocation and fluxes of whole above- and below-ground tree parts.

Portugal

Propagation of the Azorean native Morell afaya (Aiton) Wilbur (João et al., 2014).

Pereira Maria, João et al., Technical University of Lisbon, Centre for Natural Resources and Environment, Lisbon, Portugal.

The production of Morella fay a (Aiton) Wilbur plants for Azorean wildlife habitat and conservation landscaping was studied. With that purpose several germination and cutting trials were conducted and the plantlets development on different substrates were measured. In the germination trials the effect of chemical scarification, stratification, temperature and photoperiod on seeds germination characteristics were studied. In the cuttings trials semi-hardwood cuttings harvested in October and planted in substrate and softwood cuttings harvested in April and placed in aeroponic conditions, to test the effect of (0% and 0.4%) indole butyric acid on rooting were conducted. After 34 weeks, fruits' scarification under a suitable light and temperature regime enhanced the percentage of germination. The best regimes of temperature and light were: 15°C/8 h or environmental conditions (starting in October), resulting, respectively, in 23% and 22.5% germination, 134 and 126 days of mean time of germination, and 82 and 72 days for the first radicle emergence. Survival of the produced plants was superior (95%) when using the soil from the plant's habitat but plant development was superior on the mixture: BVB (NPK): perlite (2:1). Mortality of cuttings was 100%.

USA

Evaluation of Three Root Growth Potential Techniques with Tree Seedlings (Rietveld, 1989).

Rietveld, W., Forestry Sciences Laboratory, Rocky Mountain Forest and Range Experiment Station, Lincoln, NE, USA.

Two tropical forest species: Ceiba pentandra (kapok), a fast-growing tree native to South and Central America and to Western Africa, and Khaya anthotheca (African mahogany), a slower-growing tree native to Central and Eastern Africa. Growth and allometric parameters of 12-month-old saplings grown in a large-scale aeroponic system and in 50-L soil containers were compared. The main advantage of growing plants in aeroponics is that their root systems are fully accessible throughout the plant life, and can be fully recovered for harvesting.

Australia

Influence of Low Oxygen Levels in Aeroponics Chambers on Eucalypt Roots Infected with *P. cinnamomi* (Burgess et al., 1998).

Burgess, Treena et al., School of Veterinary and Life Sciences Murdoch University Murdoch, Australia.

The differences in shoot and root size were studied between the fast-growing C. pentandra and the slower-growing K. anthotheca. Roots were recovered from the aeroponically grown saplings only, and their distribution among various diameter classes followed the patterns expected from the literature. Stem, branch and leaf allometric parameters were similar for saplings of each species grown in the two systems.

Australia

Effects of Sheared-Root Inoculum of *Glomus intraradices* on Wheat Grown at Different Phosphorus Levels in the Field (Mohammad et al., 2004).

Mohammad, A. et al., School of Science, Food and Horticulture, University of Western Sydney, Australia.

An aeroponic tree growth system were studied for determining the basic allometric relationships between root and shoot components of these trees, and hence can be used to study carbon allocation and fluxes of whole above- and below-ground tree parts.

VEGETABLES

Colombia

Development of An Aeroponic System for Vegetable Production (Reyes et al., 2012).

Reyes, J.L. et al., Department of Agricultural Sciences, National University of Colombia, Bogotá, Colombia.

A study was conducted using aeroponic systems and proposed structures, ways of monitoring, watering and plant care that are flexible for the diverse types of crops so that Colombia could reduce technology dependency from overseas. Aeroponics is a way of planting in which plants are suspended on the air and grow in a humid environment without soil. This technique has most advantages when compared to other planting techniques used commonly in Mexico. With aeroponics one can control humidity, temperature, pH, and water conductivity under a greenhouse.

USA

Increased Oxygen Bioavailability Improved Vigor and Germination of Aged Vegetable Seeds (Liu et al., 2012).

Guodong Liu et al., Horticultural Sciences Department, IFAS, University of Florida, Gainesville, FL, USA.

Germination of selected vegetable seeds were studied including corn (*Zea mays* L.), squash (*Cucurbita pepo* L.), and tomato (*Solanum lycopersicum* L.) in water with different concentrations of hydrogen peroxide solution ranging from 0.06% to 3.0% (v/v) or in aeroponics, all with 0.5 mM calcium sulfate. Imbibition, oxygen consumption, proton extrusion, and alcohol dehydrogenase (ADHase) activity of corn seeds were measured gravimetrically, electrochemically, and colorimetrically as appropriate. The results showed that 0.15% hydrogen peroxide provided the optimum oxygen concentration for seed germination. The germination percentage of aged corn seeds treated with peroxide was significantly greater than those without treatment. Corn embryo orientation in relation to a moist substrate also significantly impacted oxygen bioavailability to the embryo and hence ADHase activity. The results from this research imply that consideration should be given to including oxygen fortification in seed coatings for aged seeds and for large seeds regardless of age.

WHEAT

Hungary
Lack of Active K+ Uptake in Aeroponically Grown Wheat Seedlings (Zsoldos et al., 1987).

Zsoldos, Ferenc et al at the University of Szeged, Szeged, Hungary.

A comparative study was conducted of the potassium (K$^+$) uptake and the growth of intact wheat seedlings (*Triticum aestivum* L. cv. GK Szeged) grown in 0.5 mM CaCl$_2$ solution and of seedlings grown on wet filter paper in Petri dishes under different experimental conditions. Aeroponic (AP) and hydroponic (HP) conditions brought about striking differences in the growth of the roots, whereas the shoot growth was not influenced. The dry weight of the roots was higher for the AP plants than for the HP plants. The AP grown seedlings exhibit a low rate of K$^+$ uptake, which seems to be a passive process. The effect of 2, 4–dinitrophenol (2, 4–DNP) clearly shows the absence of an active component of the K$^+$ uptake in roots grown in air with a high relative humidity. In plants grown under AP conditions the effect of Ca^{2+} on the K$^+$ uptake is unfavourable, i.e., there is an inhibition (negative Viets effect). The results relating to the effect of 2,4–DNP suggest that the "negative Viets effect" is a feature of the passive K$^+$ uptake. The data suggest that the AP growth conditions play a very important role in the induction and/or development of the ion transport system(s), which becomes impaired under the AP conditions.

Iran
Role of Scavenging Enzymes and Hydrogen Peroxide and Glutathione S-transferase in Mitigating the Salinity Effects on Wheat (Esfandiari et al., 2014).

Esfandiari, Ezatollah et al., University of Maragheh, Faculty of Agriculture, Department of Agronomy and Plant Breeding, Maragheh, Iran.

The effects of salt stress on activities of hydrogen peroxide scavenging enzymes, glutathione S-transferase, some oxidative stress markers and Na$^+$ and K$^+$ distribution patterns in sensitive (Koohdasht) and tolerant (Gaskogen) wheat varieties using aeroponics culture were studied. The seedlings were fed by nutrition solution till 3–4 leaf stage then the medium was added 200 mM NaCl. The plants were held at this condition for 14 days. The results indicated that Gaskogen had always more shoot dry matter than Koohdasht. The results obtained from this study showed that activity of scavenging enzyme like hydrogen peroxide together with glutation S-transferase caused controlling of toxic compounds in the Gaskogen variety and suppressed oxidative stress affects in compared to Kouhdasht that could refer to lower rate of hydrogen peroxidase and less lipid peroxidation in Koohdasht. As a final result, it could be stated that H$_2$O$_2$-scavenging enzymes and glutathione S-transferase had special roles in detoxification of toxic compounds leading to keep stable conditions inside the plant cells in salinity conditions. In addition, it is suggested that for better evaluation of salt tolerance in wheat genotypes or varieties and optimum utilization of genetic resources, addition to sodium rate and its allocation in various parts of the plant, defense mechanisms should be considered against oxidative stress induced salinity.

Iran

Evaluation of Cd Effects on Growth and Some Oxidative Stress Parameters of Wheat Cultivars During Seedling Stage (Esfandiari et al., 2016).

Ezatollah Esfandiari et al., University of Maragheh, Faculty of Agriculture, Department of Agronomy and Plant Breeding, Maragheh, Iran.

Six different cultivars of wheat cultivated by aeroponic system were studied. They were treated by 200 mM cadmium chloride for 14 days at the 3 or 4 leaves stage. Finally, plants were sampled and morphological characteristics, parameters subjected to defense mechanisms as well as amount of accumulated Cd in tissues were evaluated. The results showed that cadmium accumulation in tissues significantly decreased root length and dry matter of root, shoot, and whole plant. Cd accumulation also increased hydrogen peroxide amount in leaf cells of Kohdasht and Pishtase cultivars.

USA

Effects of Barley Yellow Dwarf Virus on Root and Shoot Growth of Winter Wheat Seedlings Grown in Aeroponic Culture (Hoffman et al., 1997).

Hoffman, T. K. et al., Department of Crop Sciences, University of Illinois, Urbana, USA.

Seedlings of eight soft red winter wheat (*Triticum aestivum*) cultivars were grown in an aeroponic mist box to study the effects of barley yellow dwarf virus (BYDV) on root and shoot growth and to look for differences in root and shoot growth among cultivars. The cultivars selected for the study were Caldwell, Cardinal, Clark, Howell, IL 87–2834, Tyler, and Pioneer brands 2548 and 2555. A split-plot treatment design was used, with uninfected and inoculated treatments as whole plots and cultivars as subplots. Differences among cultivars were found for most growth characteristics under both control and BYDV-infected conditions. There was a strong positive correlation between shoot and root dry weights in both the control and BYDV treatments, indicating that cultivars with vigorous shoot growth tended to have more vigorous root growth. The results indicate that, initially, the root system is affected more severely than the shoot in BYDV-infected wheat seedlings.

YAMS

Nigeria

Aeroponics: High-quality Seed Yam Production (Alawode et al., 2017).

Oluyinka Alawode, International Institute of Tropical Agriculture (IITA), Ibadan, Nigeria.

The mass production of seed yams were studied in a rapid and affordable process, which avoids using soil and transferring disease. A project in Ghana and Nigeria is using aeroponics to address the inefficiency of traditional seed yam production and increase yields of the staple crop.

Nigeria

Improved Propagation Methods to Raise the Productivity of Yam (*Dioscorea rotundata* Poir.) (Aighewi et al., 2015).

Aighewi, B. et al., Department of Crop Science, University of Abuja, Nigeria.

The white Guinea yam (*Dioscorea rotundata* Poir.) which is an important staple to millions of people in West Africa was studied. Obtaining good quality planting material for yam cultivation is a major challenge. Multiplication ratios are low, and seed tubers are prone to contamination with pests and pathogens in the traditional systems of production. New methods that have been developed to address some of the challenges of quantity and quality of seed tubers are not yet widely applied, so farmers continue to use traditional methods and save seed from a previous harvest to plant the ware crop. This document presents an overview of traditional and modern methods of seed yam production and gives a perspective for the future. Among the modern methods of seed yam production, only the minisett technique, which uses 25–100 g tuber pieces, is currently used at farmer level, although on a limited scale. While tissue and organ culture techniques are the most rapid methods of multiplying disease-free propagules, their limitations include high costs, need for skilled personnel, and specialized equipment. The aeroponics and temporary immersion bioreactor methods of producing seed yam are relatively new, and still need more research. To build and sustain a viable seed yam production system, a multiplication scheme is required that combines two or more methods including tissue culture for cleaning the seed stock.

Nigeria

Yam Production using Aeroponics Technology (Balgun et al., 2014).

Balgun, M.N. et al., Department of Crop Protection and Environmental Biology, University of Ibadan, Nigeria.

Scientists at the International Institute of Tropical Agriculture have successfully propagated yam by planning vine cuttings in Aeroponics System (AS) boxes. A loan agreement was signed with the African Development Bank (AfDB) for the Artisanal Fisheries Support Project.

REFERENCES

Abdullateef, S. et al., 2012, Potato minituber production at different plant densities using an aeroponic system, *Acta Horticulturae* 927: 53.

Abhijith, Y.C. et al., 2017, Effect of micronutrients on growth and yield of aonla (*Emblicaofficinalis gaertn.*) CV. Na-7, *Annals of Horticulture* 10(2): 176–179.

Abolitz, M. et al., 1995, Effects of pH in the root environment on leakage of phenolic compounds and mineral ions from roots of *Rosa indica* major, *Advances in Horticultural Science* 10(4): 210–214.

Aighewi, B. et al., 2015, Improved propagation methods to raise the productivity of yam (*Dioscorea rotundata Poir.*). *Food Security* 7(4): 823–834.

Al-Kodmany, K., 2018, The vertical farm: A review of developments and implications for the vertical city. *Buildings* 8(2): 24, 2075–5309.

Alawode, O. et al., 2017, *Aeroponics: High-Quality Seed Yam Production*. Ibadan, Nigeria, Spore: International Institute of Tropical Agriculture (IITA).

Albaho, M. et al., 2008, Evaluation of hydroponic techniques on growth and productivity of greenhouse grown bell pepper and strawberry. *International Journal of Vegetable Science* 14(1): 23–40.

Albornoz, F. et al., 2014, Effect of different day and night nutrient solution concentrations on growth, photosynthesis, and leaf NO_3-content of aeroponically grown lettuce. *Chilean Journal of Agricultural Research* 74(2): 240–245.

Albornoz, F. et al., 2015, Over fertilization limits lettuce productivity because of osmotic stress. *Chilean Journal of Agricultural Research* 75(3): 284–290.

Asran, M. R. et al., 2003, Pathogenicity of *Fusarium graminearum* isolates on maize (*Zea mays L.*) cultivars and relation with deoxynivalenol and ergosterol contents. *Zeitschrift für Pflanzenkrankheiten und Pflanzenschutz/Journal of Plant Diseases and Protection* 110(3): 209–219.

Baek, G. Y. et al., 2013, The effect of LED light combination on the anthocyanin expression of lettuce, *5th IFAC Conference on Bio-Robotics, IFAC Proceedings* 46(4): 120–123.

Bagyaraj, D. J., 1992, 19 Vesicular-arbuscular mycorrhiza: Application in agriculture. *Methods in Microbiology* 24: 359–373.

Baharuddin et al., 2014, An early detection of latent infection of ralstonia solanacearum on potato tubers. *International Journal of Agriculture System* 2(2): 183–188.

Balgun, M. N. et al., 2014, Yam production using aeroponic technology. *Annual Research and Review in Biology* 4 (24): 3894.

Barak, P. et al., 1996, Measurement of short-term nutrient uptake rates in cranberry by aeroponics. *Plant, Cell & Environment* 19(2): 237–242.

Barupal, Meena et al., 2018, In vitro growth profile and comparative leaf anatomy of the C3–C4 intermediate plant Mollugo nudicaulis Lam. *In Vitro Cellular & Developmental Biology Plant* 54(6): 12, 689–700.

Battaglia, D., 2017, Aeroponic gardens and their magic: Plants/persons/ethics in suspension. *History & Anthropology* 28(3): 263–292.

Biddinger et al., 1998, Physiological and molecular responses of aeroponically grown tomato plants to phosphorus deficiency. *Journal of the American Society for Horticultural Science* 123: 330–333.

Brauner, K. et al., 2015, Measuring whole plant carbon dioxide exchange with the environment reveals opposing effects of the gin2-1 mutation in shoots and roots of *Arabidopsis thaliana*, *Plant Signaling & Behavior* 10(1): e973822.

Buckseth, T. et al., 2016, Review: Methods of pre-basic seed potato production with special reference to aeroponic. *Scientia Horticulturae* 204: 79–87.

Burgess, T. et al., 1998, Influence of low oxygen levels in aeroponics chambers on eucalyptus roots infected with *Phytophthora cinnamomic*, *Plant Disease* 82(4): 6, 368–373.

Burgess, T. et al., 1999a, Effects of hypoxia on root morphology and lesion development in *Eucalyptus marginata* infected with *Phytophthora cinnamomic*. *Plant Pathology* 48(6): 786–796.

Burgess, T. et al., 1999b, Increased susceptibility of *Eucalyptus marginata* to stem infection by *Phytophthora cinnamomi* resulting from root hypoxia. *Plant Pathology* 48(6): 797–806.

Cai, M. et al., 2013, Immobilization of aluminum with mucilage secreted by root cap and root border cells is related to aluminum resistance in *Glycine max L. Environmental Science & Pollution Research* 20(12): 8924–8933.

Calori, A. H. et al., 2017, *Electrical Conductivity of the Nutrient Solution and Plant Density in Aeroponic Production of Seed Potato under Tropical Conditions.* Brazil: Instituto Agronômico de Campinas.

Chandra, S. et al., 2014, Assessment of total phenolic and flavonoid content, antioxidant properties, and yield of aeroponically and conventionally grown leafy vegetables and fruit crops: A comparative study. *Evidence-based Complementary & Alternative Medicine (eCAM)* 2014: 1–9.

Chandran, R. et al., 2009, Arbuscular mycorrhizal fungal inoculum production using *Ipomoea batata* hairy roots in bioreactor. *ICFAI Journal of Biotechnology* 3(2): 56–64.

Chang, D. et al., 2008, Physiological growth responses by nutrient interruption in aeroponically grown potatoes. *American Journal of Potato Research* 85(5): 315–323.

Chang, D. et al., 2011, Growth and yield response of three aeroponically grown potato cultivars (*Solanum tuberosum L.*) to different electrical conductivities of nutrient solution. *American Journal of Potato Research* 88(6): 450–458.

Chang, D. et al., 2012, Growth and tuberization of hydroponically grown potatoes. *Potato Research* 55(1): 69–81.

Chang, D. et al., 2016, Nutritional and structural response of potato plants to reduced nitrogen supply in nutrient solution. *American Journal of Potato Research* 93(4): 368–377.

Chen, W. et al., 2008, Effects of aluminum (+3) on the biological characteristics of cowpea root border cells. *Acta Physiologiae Plantarum* 30(3): 303–308.

Chen, Y. et al., 2017, Pyramid shaped hydroponic and aeroponic technology – a new technology for pepper cultivation. *Agricultural Science & Technology* 18(3): 521–523.

Chica Toro, F. D. J. et al., 2018, Absorption curves – mineral-extraction under an aeroponic system for white chrysanthemum (*Dendranthema grandiflorum* (Ramat.) Kitam. cv. Atlantis White). *Acta Agronomica* 67(1): 86–93.

Chipanthenga, M. et al., 2013, Performance of different potato genotypes under aeroponics system. *Journal of Applied Horticulture* 15(2): 142–146.

Correll, M. J. et al., 2001, One-step acclimatization of plantlets using a mist reactor. *Biotechnology & Bioengineering* 73(3): 253–258.

da Silva, F. et al., 2018, Evaluation of "UFV aeroponic system" to produce basic potato seed minitubers. *American Journal of Potato Research* 95(5): 443–450.

Dolven, B., 1998, Rooting for lettuce: Aero-Green Technology, Singapore: Growing vegetables aeroponically – or without soil. *Far Eastern Economic Review* 161(43): 48.

du Toit, L. J. et al., 1997, Evaluation of an aeroponics system to screen maize genotypes for resistance to *Fusarium graminearum* seedling blight. *Plant Disease* 81: 175–179.

El-Behairy, U. A. et al., 2003, Effect of side and level of cultivation on production and quality of strawberry produced by aeroponic system. *Acta horticulturae* 608: 43–51.

Engenhart, M., 1984, The influence of lead ions on the productivity and content of minerals in *Phaseolus vulgaris L.* in hydroponics and aeroponics. *Flora – Morphology Distribution Functional Ecology of Plants* 175(4): 273–282.

eQiao, Z. et al., 2013, Unleashing the potential of the root hair cell as a single plant cell type model in root systems biology. *Frontiers in Plant Science* 4: 484.

Esfandiari, E. et al., 2014, Role of scavenging enzymes and hydrogen peroxide and glutathione S-transferase in mitigating the salinity effects on wheat. *Journal of Plant Biology* 6(20): preceding 1, 2008–8264.

Esfandiari, E. et al., 2016, Evaluation of Cd effects on growth and some oxidative stress parameters of wheat cultivars during seedling stage. *Zīst/shināsī-i Giyāhī-i Īrān* 8(27): 1–16.

Eshel, A., Waisel, Y., (1997) Aeroponics. In: Altman A., Waisel Y. (eds) *Biology of Root Formation and Development. Basic Life Sciences*, Vol. 65. Berlin: Springer.

Eshel, A. et al., 2001, Allometric relationships in young seedlings of faba bean (*Vicia faba L.*) following removal of certain root types. *Plant and Soil* 233(2): 161–166.

Eshel, A. et al., 2005, Roots of tomato respond to mechanical stimulation by induction of touch genes. *Plant Biosystems* 139(2): 209–213.

Eshel, A. et al., 2013, Root–shoot allometry of tropical forest trees determined in a large-scale aeroponic system. *Annals of Botany* 112(2): 291–296.

Evans, L. S. et al., 1980, Effect of nutrient medium pH on symbiotic nitrogen fixation by *Rhizobium leguminosarum* and *Pisum sativum*. *Plant and Soil* 56(1): 71–80.

Everett, K. T. et al., 2010, Douglas-fir seedling response to a range of ammonium nitrate ratios in aeroponic culture. *Journal of Plant Nutrition* 33(11): 1638–1657.

Factor, T. et al., 2007, Potato basic minitubers production in three hydroponic systems. *Horticultura Brasileira* 25(1): 82–87.

Farissi, M. et al., 2018, Variations in leaf gas exchange, chlorophyll fluorescence and membrane potential of *Medicago sativa* root cortex cells exposed to increased salinity: The role of the antioxidant potential in salt tolerance. *Archives of Biological Sciences* 70(3): 413–423.

Farran, I. et al., 2006, Potato minituber production using aeroponics: Effect of plant density and harvesting intervals. *American Journal of Potato Research* 83(1): 47–53.

Finlayson, S. A. et al., 1996a, Localization of ethylene biosynthesis in roots of sunflower (*Helianthus annuus*) seedlings. *Physiologia Plantarum* 96(1): 36–42.

Finlayson, S. A. et al., 1996b, The effect of carbon dioxide on ethylene evolution and elongation rate in roots of sunflower (*Helianthus annuus*) seedlings. *Physiologia Plantarum* 98(4): 875–881.

Fira, A. et al., 2012, Direct ex vitro rooting and acclimation in Blackberry Cultivar 'Loch Ness', Bulletin of the University of Agricultural Sciences & Veterinary Medicine Cluj-Napoca. *Animal Science & Biotechnologies* 69(1/2): 247–254.

Fischinger, S. A. et al., 2010, Elevated carbon dioxide concentration around alfalfa nodules increases nitrogen fixation. *Journal of Experimental Botany* 61(1): 121–130.

Freundl, E. et al., 1998, Apoplastic transport of abscisic acid through roots of maize: Effect of the exodermis. *Planta* 206(1): 7–19.

Garrido, I., 1998, Effect of some electron donors and acceptors on redox capacity and simultaneous net H+/K+ fluxes by aeroponic sunflower seedling roots: Evidence for a CN-resistant redox chain accessible to nonpermeative redox compounds. *Protoplasma*, 205(1–4): 141–155.

Garrido, I., 2003, Redox-related peroxidative responses evoked by methyl-jasmonate in axenically cultured aeroponic sunflower (*Helianthus annuus L.*) seedling roots. *Protoplasma* 221 (1–2): 79–91.

Garrido, I. et al., 1998, Net simultaneous hydrogen and potassium ion flux kinetics in sterile aeroponic sunflower seedling roots: Effects of potassium ion supply, valinomycin, and dicyclohexylcarbodiimide. *Journal of Plant Nutrition* 21(1): 115–137.

Gąsecka, M. et al., 2009, The effect of temperature and crown size on asparagus yielding. *Folia Horticulturae* 21(2), 49–59.

Gaudin, A. C. M. et al., 2011, Novel temporal, fine-scale and growth variation phenotypes in roots of adult-stage maize (*Zea mays L.*) in response to low nitrogen stress. *Plant, Cell & Environment* 34(12): 2122–2137.

Gaudin, A. C. M. et al., 2014, The effect of altered dosage of a mutant allele of Teosinte branched 1 (tb1-ref) on the root system of modern maize. *BMC Genetics* 15(1): 1–28, 28.

Geier, T. et al., 2008, Production and rooting behaviour of rol B-transgenic plants of grape rootstock 'Richter 110' (*Vitis berlandieri×V. rupestris*). *Plant Cell, Tissue & Organ Culture* 94(3): 269–280.

Gianinazzi, S. et al., 2004, Inoculum of arbuscular mycorrhizal fungi for production systems: Science meets business. *Canadian Journal of Botany* 82(8): 1264–1271.

Giurgiu, R. M. et al., 2017, A study of the cultivation of medicinal plants in hydroponic and aeroponic technologies in a protected environment. *Hortic* 1170, 671–678.

Gregory, P. J. et al., 2009, Root phenomics of crops: Opportunities and challenges. *Functional Plant Biology* 36(10/11): 922–929.

Groves, E. et al., 2015, Role of salicylic acid in phosphite-induced protection against Oomycetes; a *Phytophthora cinnamomi* – *Lupinus augustifolius* model system. *European Journal of Plant Pathology* 141(3): 559–569.

Gwynn-Jones, D. et al., 2018, Can the optimisation of pop-up agriculture in remote communities help feed the world? *Global Food Security* 18: 35–43.

Hachez, C. et al., 2012, Short-term control of maize cell and root water permeability through plasma membrane aquaporin isoforms. *Plant, Cell & Environment* 35(1): 185–198.

Hartmann, A. et al., 2018, Implementation and application of a root growth module in HYDRUS. *Vadose Zone Journal* 17(1): 7–7.

Hartung, W. et al., 2002, Abscisic acid concentration, root pH and anatomy do not explain growth differences of chickpea (*Cicer areitinum L.*) and lupin (*Lupinus angustifolius L.*) on acid and alkaline soils. *Plant & Soil* 240(1): 191.

Hassanpanah, D., 2014, Evaluating potential production of mid-late maturing minituber of potato cultivars and promising clones under aeroponic system. *Ikuṭīziyuluzhī-i Giyāhān-i Zirāī* 8(3(31)): 331–346.

Hawes, M. C. et al., 1990, Correlation of pectolytic enzyme activity with the programmed release of cells from root caps of pea (*Pisum sativum*). *Plant Physiology* 94(4): 1855–1859.

Hawkins, B. J., 2007, Family variation in nutritional and growth traits in Douglas-fir seedlings. *Tree Physiology* 27(6): 911–919.

Hayden, A. L., 2006, Aeroponic and hydroponic systems for medicinal herb, rhizome, and root crops. *HortScience* 41(3): 536–538.

Hays, S. M., 1993, Grenzwurzeln? *Agricultural Research* 41(8): 10.

He, J., 2015, Farming of vegetables in space-limited environments. *COSMOS* 11(1): 16, 21–36.

He, J. et al., 2008, Interaction between iron stress and root-zone temperature on physiological aspects of aeroponically grown Chinese broccoli. *Journal of Plant Nutrition* 31(1/3): 173–192.

He, J. et al., 2010, Effects of elevated root zone carbon dioxide and air temperature on photosynthetic gas exchange, nitrate uptake, and total reduced nitrogen content in aeroponically grown lettuce plants. *Journal of Experimental Botany* 61(14): 3959–3969.

He, J. et al., 2013, Impact of climate change on food security and proposed solutions for the modern city. *Acta Hort* 1004: 3.

He, J. et al., 2015, Growth irradiance effects on productivity, photosynthesis, nitrate accumulation and assimilation of aeroponically grown brassica alboglabra. *Journal of Plant Nutrition* 38(7): 1022–1035.

Henry, A., 2013, IRRI's drought stress research in rice with emphasis on roots: Accomplishments over the last 50 years. *Plant Root* 7: 92–106.

Henzler, T. et al., 1999, Diurnal variations was conducted in hydraulic conductivity and root pressure can be correlated with the expression of putative aquaporins in the roots of *Lotus japonicas*. *Planta* 210(1): 50–60.

Hikosaka, Y. et al., 2015, Dry-fog aeroponics affects the root growth of leaf lettuce (*Lactuca sativa L.* cv. *Greenspan*) by changing the flow rate of spray fertigation. *Environmental Control in Biology* 53(4): 181–187.

Hoffman, T. K. et al., 1997, Effects of barley yellow dwarf virus on root and shoot growth of winter wheat seedlings grown in aeroponic culture. *Plant Disease* 81: 497–500.

Hoyos, García et al., 2012, Growing degree days accumulation in a cucumber (*Cucumis sativus L.*) crop grown in an aeroponic production model. *Revista Facultad Nacional de Agronomía Medellín* 65(1): 6389–6398.

Hung, L-L. L. et al., 1988, Production of vesicular-arbuscular mycorrhizal fungus inoculum in aeroponic culture. *Applied & Environmental Microbiology* 54: 353–357.

Hung, L-L. L. et al., 1991, Use of hydrogel as a sticking agent and carrier for vesicular-arbuscular mycorrhizal fungi. *Mycological Research* 95(4): 427–429.

Hussein A. et al., 2018, Mini tuber production in potato via aeroponic system. *Türkiye Tarımsal Araştırmalar Dergisi* 5(1): 79–85.

Ijdo, M. et al., 2011, Methods for large-scale production of AM fungi: Past, present, and future. *Mycorrhiza* 21(1): 1–16.

Jackson, T. J. et al., 2000, Action of the fungicide phosphite on *Eucalyptus marginata* inoculated with Phytophthora cinnamomic. *Plant Pathology* 49(1): 147–154.

Jarstfer, A. G. et al., 1998, Tissue magnesium and calcium affect arbuscular mycorrhiza development and fungal reproduction. *Mycorrhiza* 7(5): 237–242.

Jianmin, G. et al., 2015, Prediction model for fundamental frequency of low frequency bending vibration piezoelectric vibrator of ultrasonic atomizing nozzle. *Transactions of the Chinese Society of Agricultural Engineering* 31(4): 55–62.

Jie, H. et al., 1998, Growth and photosynthetic characteristics of lettuce (*Lactuca sativa L.*) under fluctuating hot ambient temperatures with the manipulation of cool root-zone temperature. *Journal of Plant Physiology* 152(4): 387–391.

João, M. et al., 2014, Propagation of the Azorean native Morell afaya (Aiton) Wilbur, Pereira. *Silva Lusitana* 22: 49–62.

Johnson, M. et al., 2015, Evaluation of algal biomass production on vertical aeroponic substrates. *Algal Research* 10: 240–248.

Jones Jr., J. B., 2005, *Hydroponics: A Practical Guide for the Soil-less Grower.* Boca Raton, FL: CRC Press, 7.

Jurga, A. et al., 2018a, Concept of experimental platform to investigate aeroponic systems in microgravity conditions. *E3S Web of Conferences* 44: 00061.

Jurga, A. et al., 2018b, Concept of aeroponic biomass cultivation and biological wastewater treatment system in extraterrestrial human base. *E3S Web of Conferences* 44: 00060.

Kacjan-Mar, N. et al., 2002, Nitrate content in lettuce (*Lactuca sativa L.*) grown on aeroponics with different quantities of nitrogen in the nutrient solution. *Acta Agronomica Hungarica* 50(4): 389–397.

Kamies, R. et al., 2010, The use of aeroponics to investigate antioxidant activity in the roots of *Xerophyta viscosa*. *Plant Growth Regulation* 62(3): 203–211.

Katin, O. et al., 2017, Optimization of the automated colorimetric measurement system for pH of liquid. *MATEC Web of Conferences* 132: 04010.

Kaur, R. P. et al., 2016, Effect of supporting medium on photoautotrophic microplant survival and growth of potato (*Solanum tuberosum*). *Current Advances in Agricultural Sciences (An International Journal)* 8(2): 172–176.

Kratsch, H. A. et al., 2006, Aeroponic system for control of root-zone atmosphere. *Environmental & Experimental Botany* 55(1/2): 70–76.

Kumar, B. et al., 2016, Hydro and aeroponic technique for rapid drought tolerance screening in maize (*Zea mays*). *Indian Journal of Agronomy* 61(4): 509–511.

Kumari, A. et al., 2016, The changes in morphogenesis and bioactivity of *Tetradenia riparia*, *Mondia whitei* and *Cyanoptis speciosa* by an aeroponic system. *Industrial Crops & Products* 84: 199–204.

Lakhiar, I. A. et al., 2018a, Experimental study of ultrasonic atomizer effects on values of EC and pH of nutrient solution. *International Journal of Agricultural & Biological Engineering* 11(5): 59–64.

Lakhiar, I. A. et al., 2018b, Modern plant cultivation technologies in agriculture under controlled environment: A review on aeroponics. *Journal of Plant Interactions* 13(1): 338–352.

Lee, S. et al., 2015, Beneficial bacteria and fungi in hydroponic systems: Types and characteristics of hydroponic food production methods. *Scientia Horticulturae* 195, 206–215.

Lemma, T. et al., 2017, Determination of nutrient solutions for potato (*Solanum tuberosum L.*) seed production under aeroponics production system. *Open Agriculture* 2(1): 155–159.

Li, Q. et al., 2018, Growth responses and root characteristics of lettuce grown in aeroponics, hydroponics, and substrate culture. *Horticulturae* 4(4): 35.

Li, Z. et al., 2014, Study of new technology supporting design of home use plant factory. *The Science of Design Bulletin of JSSD* 61(3): 3_45–3_50.

Li, Z. et al., 2015, Protein dynamics in young maize root hairs in response to macro- and micronutrient deprivation. *Journal of Proteome Research* 14(8): 3362.

Liu, G. et al., 2012, Increased oxygen bioavailability improved vigor and germination of aged vegetable seeds. *Hort Science* 47: 1714–1721.

Liu, Y.-L. et al., 2010, Impacts of root-zone hypoxia stress on muskmelon growth, its root respiratory metabolism, and antioxidative enzyme activities. *Yingyong Shengtai Xuebao* 21(6): 1439–1445.

Liu, Y.-L. et al., 2013, Effects of elevated rhizosphere carbon dioxide concentration on the photosynthetic characteristics, yield, and quality of muskmelon. *Yingyong Shengtai Xuebao* 24(10): 2871–2877.

Lovy, L. et al., 2013, Cadmium uptake and partitioning in the hyperaccumulator *Noccaea caerulescens* exposed to constant Cd concentrations throughout complete growth cycles. *Plant & Soil* 362(1/2): 345–355.

Maggi, F. et al., 2010a, Martian base agriculture: The effect of low gravity on water flow, nutrient cycles, and microbial biomass dynamics. *Advances in Space Research* 46(10): 1257–1265.

Maggi, F. et al., 2010b, Space agriculture in micro- and hypo-gravity: A comparative study of soil hydraulics and biogeochemistry in a cropping unit on Earth, Mars, the Moon and the space station. *Planetary and Space Science* 58(14): 1996–2007.

Maršić, N. K. et al., 2002, Effects of different nitrogen levels on lettuce growth and nitrate accumulation in iceberg lettuce (*Lactuca sativa var. capitata L.*) grown hydroponically under greenhouse conditions. *Die Gartenbauwissenschaft* 67(4): 128–134.

Martin-Laurent, F. et al., 1997, A new approach to enhance growth and nodulation of *Acacia mangium* through aeroponic culture. *Biology & Fertility of Soils* 25(1): 6, 7–12.

Martin-Laurent, F. et al., 1999, The aeroponic production of *Acacia mangium* saplings inoculated with AM fungi for reforestation in the tropics. *Forest Ecology and Management* 122(3): 199–207.

Mateus-Rodruguez, J. et al., 2012, Response of three potato cultivars grown in a novel aeroponics system for mini-tuber seed production. *Acta Hortic* 947: 46.

Mateus-Rodriguez, J. R. et al., 2013, Technical and economic analysis of aeroponics and other systems for potato mini-tuber production in Latin America. *American Journal of Potato Research* 90(4): 12, 357–368.

Mateus-Rodriguez, J. F. et al., 2014, Genotype by environment effects on potato mini-tuber seed production in an aeroponics system. *Agronomy* 4(4): 514–528.

Mattson, N. S. et al., 2012, Effect of light regimen on yield and flavonoid content of warehouse grown Aeroponic *Eruca sativa*. *Acta Hortic* 956: 49.

Mehandru, P. et al., 2014, Evaluation of aeroponics for clonal propagation of *Caralluma edulis, Leptadenia reticulata* and *Tylophora indica* – three threatened medicinal Asclepiads. *Physiology & Molecular Biology of Plants* 20(3): 365–373.

Mehra, M. et al., 2018, Original papers: IoT based hydroponics system using deep neural networks. *Computers and Electronics in Agriculture* 155: 473–486.

Meyer, C. J. et al., 2009, Environmental effects on the maturation of the endodermis and multiseriate exodermis of Iris germanica roots. *Annals of Botany* 103(5): 687–702.

Miller, M., 2020, Aeroponics: A sustainable solution for urban agriculture. Environmental Law Institute, ELI.org, Vibrant Environment Blog, 4/4/18.

Miyamoto, N. et al., 2001, Hydraulic conductivity of rice roots. *Journal of Experimental Botany* 52(362): 1835–1846.

Mohammad, A. et al., 2000, Improved aeroponic culture of inocula of arbuscular mycorrhizal fungi. *Mycorrhiza* 9(6): 337–339.

Mohammad, A. et al., 2004, Effects of sheared-root inoculum of *Glomus intraradices* on wheat grown at different phosphorus levels in the field. *Agriculture, Ecosystems & Environment* 103(1): 245–249.

Montoya, A. P. et al., 2017, Automatic aeroponic irrigation system based on Arduino's platform. *Journal of Physics: Conference Series* 850(1): 1.

Morris, J. B., 2014, Seed production from aeschynomene genetic resources rescued and regenerated using aeroponics. *Seed Technology* 36(2): 115–122.

Mueller, D. S. et al., 2002, Use of aeroponic chambers and grafting to study partial resistance to *Fusarium solani* f. sp. glycines in soybean. *Plant Disease* 86(11): 1223–1226.

Muthoni, J. et al., 2013, Alleviating potato seed tuber shortage in developing countries: Potential of true potato seeds. *Australian Journal of Crop Science* 7(12): 1946–1954.

Muthoni, J. et al., 2014, Multiplication of seed potatoes in a conventional potato breeding programme: A case of Kenya's national potato programme. *Australian Journal of Crop Science* 8(8): 1195–1199.

Muthuraj, R. et al., 2016, Effect of Micro-plants hardening on aeroponic potato seed production. *Potato Journal* 43(2): 214–219.

Nguyen, T. K. O. et al., 2013, From bioreactor to entire plants: Development of production systems for secondary metabolites. *Advances in Botanical Research* 68: 205–232.

Nichols, M. A. et al., 2002, Continuous production of greenhouse crops using aeroponics. *Acta Horticulturae* 578: 289–291.

Nordborg, F. et al., 2001, Growth responses of rooted cuttings from five clones of *Picea abies (L.) Karst.* after a short drought period. *Scandinavian Journal of Forest Research* 16(4): 324–330.

Nurwahkyuningsih et al., 2013, Aplikasi zone cooling pada sistem aeroponik kentang di daratan medium tropika basah. *Journal Keteknikan Pertanian* 1(1): 2013.

Nurwahyuningsih et al., 2017, Aplikasi root zone cooling system untuk perbaikan pembentukan umbi bawang merah (*Allium cepa var. aggregatum*). *Journal Keteknikan Pertanian* 5(2): 107–112.

Oakes, A. D. et al., 2012, Vegetative propagation of American Elm (*Ulmus americana*) varieties from softwood cuttings. *Journal of Environmental Horticulture* 30(2): 73–76.

Oraby, H. et al., 2015, A low nutrient solution temperature and the application of stress treatments increase potato mini-tubers production in an aeroponic system. *American Journal of Potato Research* 92(3): 387–397.

Ottosson, B. et al., 1997, Feature article: Transpiration rate in relation to root and leaf growth in cuttings of Begonia X hiemalis. *Scientia Horticulturae* 68(1): 125–136.

Padgett, P. E. et al., 1993, Contamination of ammonium-based nutrient solutions by nitrifying organisms and the conversion of ammonium to nitrate. *Plant Physiology* 101(1): 141–146.

Pagliarulo, C. et al., 2004, Potential for greenhouse aeroponic cultivation of *Urtica dioica*. *Acta Hort* 659: 61–66.

Pan, J-W. et al., 2004, Root border cell development is a temperature-insensitive and al-sensitive process in barley. *Plant & Cell Physiology* 45(6): 751–760.

Pellerin, S. et al., 1995, Length of the apical unbranched zone of maize axile roots: Its relationship to root elongation rate. *Environmental and Experimental Botany* 35: 193–200.

Peng, C. et al., 2016, Distribution and speciation of cu in the root border cells of rice by STXM combined with NEXAFS. *Bulletin of Environmental Contamination & Toxicology* 96(3): 408–414.

Peterson, L. A. et al., 1988, An intermittent aeroponics system. *Crop Science* 28: 712–713.

Peterson, L. A. et al., 1991, An intermittent aeroponics system adaptable to root research. *Crop Science* 24, 628–631.

Prastowo et al., 2007, Irrigation efficiency and uniformity of aeroponics system a case study in parung hydroponics farm. *Journal Keteknikan Pertanian* 21(2): 127–134.

Pratiwi, P. R. et al., 2015, Pengaruh tingkat EC (electrical conductivity) terhadap Pertumbuhan Tanaman Sawi (*Brassica juncea L.*) pada Sistem Instalasi Aeroponik Vertikal. *Jurnal Agro* 0(1): 50–55.

Rahaim, C. P. et al., 2008, A greenhouse for mars and beyond. *AIP Conference Proceedings* 969(1): 917–924.

Rao, A. et al., 1995, Aeroponics chambers for evaluating resistance to Aphanomyces root rot of peas (*Pisum sativum*). *Plant Disease* 79: 128–132.

Rattan, L., 2016, Feeding 11 billion on 0.5 billion hectare of area under cereal crops. *Food & Energy Security* 35(4): 239–251.

Redjala, T. et al., 2011, Relationship between root structure and root cadmium uptake in maize. *Environmental and Experimental Botany* 71(2): 241–248.

Reissinger, A. et al., 2003, Infection of barley roots by *Chaetomium globosum*: Evidence for a protective role of the exodermis. *Mycological Research* 107(9): 1094–1102.

Reyes, J. L. et al., 2012, Development of an aeroponic system for vegetable production. *Acta Hortic* 947: 18.

Rietveld, W., 1989, Evaluation of three root growth potential techniques with tree seedlings. *New Forests* 3(2): 181–189.

Ritter, E. et al., 2001, Comparison of hydroponic and aeroponic cultivation systems for the production of potato minitubers. *Potato Research* 44(2): 127–135.

Robertson, J. M. et al., 1985, Drought-induced increases in abscisic acid levels in the root apex of sunflower. *Plant Physiology* 79(4): 1086–1089.

Ronzón-Ortega, M. et al., 2015, PRODUCCIÓN ACUAPÓNICA DE TRES HORTALIZAS EN SISTEMAS ASOCIADOS AL CULTIVO SEMI-INTENSIVO DE TILAPIA GRIS (*Oreochromis niloticus*). *Agroproductividad* 8(3): 26–32.

Roper, T. R. et al., 2004, Rate of ammonium uptake by cranberry (*Vaccinium macrocarpon Ait.*) vines in the field is affected by temperature. *HortScience* 39(3): 588–590.

Rossi, L. et al., 2015, Salt stress modifies apoplastic barriers in olive (*Olea europaea L.*): A comparison between a salt-tolerant and a salt-sensitive cultivar. *Scientia Horticulturae* 31(192): 38–46.

Salachasa, G. et al., 2015, Yield and nutritional quality of aeroponically cultivated basil as affected by the available root-zone volume. *Emirates Journal of Food & Agriculture (EJFA)* 27(12): 911–918.

Sang, Y. et al., 2014, Study on optimization of hydroponic technology of virus-free potato plantlets in winter in chengdu plain. *Agricultural Science & Technology* 15(12): 2096–2099.

Sani, M. I. et al., 2016, Web-based monitoring and control system for aeroponics growing chamber, *2016 International Conference on Control, Electronics, Renewable Energy and Communications (ICCEREC) Control*, 162–168.

Santos, K. M. et al., 2009, Stem versus Foliar Uptake during Propagation of Petunia Xhybrida Vegetative Cuttings. *HortScience* 44(7): 1974–1977.

Saraswati, I. et al., 2018, Applications of temperature and humidity monitoring system at aerophonic plants based on IoT. *MATEC Web of Conferences* 218: 03017.

Schurr, U. et al., 2001, Dynamics of concentrations and nutrient fluxes in the xylem of *Ricinus communis* – diurnal course, impact of nutrient availability and nutrient uptake. *Plant, Cell & Environment* 24(1): 41–52.

Sharaf-Eldin, M. A. et al., 2006, Movement and containment of microbial contamination in the nutrient mist bioreactor. *In Vitro Cellular & Developmental Biology Plant* 42(6): 553–557.

Sharma, U. et al., 2018, Aeroponics for adventitious rhizogenesis in evergreen haloxeric tree & *Tamarix aphylla (L.) Karst.*: Influence of exogenous auxins and cutting type. *Physiology & Molecular Biology of Plants* 24(1): 167–174.

Silberbush, M. et al., 1988, Response of peanuts (*Arachis hypogaea L.*) grown in saline nutrient solution to potassium nitrate. *Journal of Plant Physiology* 132(2): 229–233.

Singh et al., 2017, Feasibility studies for two consecutive crops of potato (*Solanum tuberosum L.*) under aeroponic system in north-western plains of India. *Research on Crops* 18(2): 264–269.

Souret, F. F. et al., 2000, The growth of saffron (*Crocus sativus L.*) in aeroponics and hydroponics. *Journal of Herbs, Spices & Medicinal Plants* 7(3): 25.

Srihajong, N. et al., 2006, Heat pipe as a cooling mechanism in an aeroponic system. *Applied Thermal Engineering* 26: 267–276.

Stadt, K. J. et al., 1992, Control of relative growth rate by application of the relative addition rate technique to a traditional solution culture system. *Plant and Soil* 142(1): 113–122.

Steen, H., 2016, Food security. *Science* 352(6288): 889–889.

Sternberg, M., 2016, From America to the holy land: Disentangling plant traits of the invasive *Heterotheca subaxillaris (Lam.)*. *Plant Ecology* 217(11): 1307–1314.

Subramaniam, R. et al., 2011, Aeroponics: Experiences from Singapore on a green technology for urban farming. Chapter 69 in *Environmental Leadership: A Reference Handbook*. Thousand Oaks, CA: Sage.

Sumarni, E. et al., 2013, Seed potato production using aeroponics system with zone cooling in wet tropical lowlands. *Acta Hortic* 1011: 16.

Sumarni, E. et al., 2016, G0 seed potential of the aeroponics potatoes seed in the lowlands with a root zone cooling into G1 in the highlands. *Rona Teknik Pertanian* 9(1): 1–10.

Sumarni, E. et al., 2019, *AgricEngInt: CIGR Journal* 21(2). Open access at www.cigrjournal.org.

Sun, Z. et al., 2011, Effects of elevated CO applied to potato roots on the anatomy and ultrastructure of leaves. *Biologia Plantarum* 55(4): 675–680.

Tabatabaei, S. J., 2008, Effects of cultivation systems on the growth and essential oil content and composition of valerian. *Journal of Herbs, Spices & Medicinal Plants* 14(1/2): 54–67.

Takahashi, H. et al., 1999, A spaceflight experiment for the study of gravimorphogenesis and hydrotropism in cucumber seedlings. *Journal of Plant Research* 112(4): 497–505.

Takano, T. et al., 1987, Trickle water and feeding system in plant culture and light-dark cycle effects on plant growth. *Advances in Space Research* 7(4): 149–152.

Tan, L. et al., 2002, Effects of root-zone temperature on the root development and nutrient uptake of *Lactuca Sativa L* "panama" grown in an aeroponic system in the tropics. *Journal of Plant Nutrition* 25(2): 297–314.

Tattini, M., 1994, Ionic relations of aeroponically-grown olive genotypes, during salt stress, *Plant & Soil* 161(2): 251–256.

Teng, Y. et al., 2014, Development of droplets penetrating roots performance test device and tests applied this device in ultrasonic aeroponic system. *Key Applied Mechanics & Materials* (680): 288–291.

Tham, F.-Y. et al., 1999, Aeroponic production of *Acacia mangium* saplings inoculated with AM fungi for reforestation in the tropics. *Forest Ecology and Management* 122(3): 199–207.

Thangavelu, M., 2014, Planet moon: The future of astronaut activity and settlement. *Architectural Design* 84(6): 1, 20–29.

Tibbitts, T. W. et al., 1994, Solid matrix and liquid culture procedures for growth of potatoes. *Advances in Space Research* 14(11): 427–433.

Tierno, R. et al., 2014, Differential growth response and minituber production of three potato cultivars under aeroponics and greenhouse bed culture. *American Journal of Potato Research* 91(4): 346–353.

Touraine, B. et al., 1988, Charge balance in NO_3–fed soybean. Estimation of K+ and carboxylate recirculation. *Plant Physiology* 88: 605–612.

Towler, M. J. et al., 2007, Using an aerosol deposition model to increase hairy root growth in a mist reactor. *Biotechnology & Bioengineering* 96(5): 881–891.

Varney, G. T. et al., 1993, Sites of entry of water into the symplast of maize roots. *New Phytologist* 125(4): 733–741.

Vaughan, M. M. et al., 2011, An aeroponic culture system for the study of root herbivory on *Arabidopsis thaliana*. *Plant Methods* 7(1): 5.

von Bieberstein, P. et al., 2014, Biomass production and withaferin a synthesis by *Withania somnifera* grown in aeroponics and hydroponics. *HortScience* 49(12): 4, 1506–1509.

Wagner, R. E. et al., 1991, Detoxification and evaluation of foam supports for aeroponically grown soybean. *Crop Science* 31: 1071–1073.

Wagner, R. E. et al., 1992, An aeroponics system for investigating disease development on soybean taproots infected with *Phytophthora sojae*. *Plant Disease* 76: 610–614.

Waisel, Y. et al., 1987, Differences in responses of various radish roots to salinity. *Plant and Soil* 104(2): 191–194.

Wang, C-C. et al., 2018, Manipulating aeroponically grown potatoes with gibberellins and calcium nitrate. *American Journal of Potato Research* 95(4): 351–361.

Weathers, P. J. et al., 1992, Research review paper: Aeroponics for the culture of organisms, tissues and cells. *Biotechnology Advances* 10(1): 93–115.

Weathers, P. J. et al., 1988, Regeneration of plants using nutrient mist culture. *In Vitro Cellular & Developmental Biology* 24(7): 727–732.

Weathers, P. J. et al., 2001, Transformed roots of *Artemisia annua* exhibit an unusual pattern of border cell release. *In Vitro Cellular & Developmental Biology. Plant* 37(4): 440–445.

Weber, J. et al., 2007, Effects of nitrogen source on the growth and nodulation of *Acacia mangium* in aeroponic culture. *Journal of Tropical Forest Science* 19(2): 103–112.

Weber, J. W. et al., 2005, Survival and growth of *Acacia mangium* willd. Bare-root seedlings after storage and transfer from aeroponic culture to the field. *Annals of Forest Science (EDP Sciences)* 62(5): 475–477.

Wu, C-G. et al., 1995, Spore Development of *Entrophospora kentinensis* in an aeroponic system. *Mycologia* 87(5): 582–587.

Xing, C-H. et al., 2008, Developmental characteristics and response to iron toxicity of root border cells in rice seedlings. *Journal of Zhejiang University: Science B* 9(3): 261–264.

Xu, Y-M. et al., 2009, 2,3-Dihydrowithaferin A-3β-O-sulfate, a new potential prodrug of withaferin A from aeroponically grown Withania somnifera, *Special Issue: Natural Products in Medicinal Chemistry, Bioorganic & Medicinal Chemistry* 17(6): 2210–2214.

Xu, Y-M. et al., 2011, Unusual withanolides from aeroponically grown *Withania somnifera*. *Phytochemistry* 72(6): 518–522.

Xu, Y-M. et al., 2016, 17β-Hydroxy-18-acetoxywithanolides from aeroponically grown *Physalis crassifolia* and their potent and selective cytotoxicity for prostate cancer cells. *Journal of Natural Products* 79(4): 821.

Xu, Y-M. et al., 2018, Cytotoxic and other withanolides from aeroponically grown *Physalis philadelphica*. *Phytochemistry* 152: 174–181.

Yi, H. et al., 2012, Interaction between Potassium concentration and Root-Zone Temperature on Growth and Photosynthesis of Temperate Lettuce Grown in the Tropics, Luo. *Journal of Plant Nutrition* 35(7): 1004–1021.

Yoojeong, K. et al., 2001, Secondary metabolism of hairy root cultures in bioreactors. *In Vitro Cellular & Developmental Biology. Plant* 38(1): 1–10.

Yue, H. et al., 2015, Effects of different cultivation patterns on nutrient and safety qualities of vegetables: A review. *Journal of Shanghai Normal University (Natural Sciences)* 44(6): 672–680.

Zdyb, A. et al., 2018, Allene oxide synthase, allene oxide cyclase and jasmonic acid levels in Lotusnodules. *PLoS ONE* 13(1): e0190884.

Zervoudakis, G. et al., 2015, Nitrogen nutrition effect on aeroponic basil (*Ocimum basilicum L.*) catalase and lipid peroxidation. *Notulae Botanicae Horti Agrobotanici Cluj-Napoca* 43(2): 561–567.

Zhao, X. et al., 2010, Effects of substrate-aeration cultivation pattern on tomato growth. *Journal of Applied Ecology* 21: 74–78.

Zimmermann, H. et al., 1998, Apoplastic transport across young maize roots: Effect of the exodermis. *Planta* 206(1): 7–19.

Zimmermann, H. M. et al., 2000, Chemical composition of apoplastic transport barriers in relation to radial hydraulic conductivity of corn roots (*Zea mays L.*), *Planta* 210(2): 302–311.

Zobel, R. W. et al., 1975, Method for growing plants aeroponically. *Plant Physiology* 57(3): 344–346.

Zou, T. et al., 2017, Responses of *Polygonatum odoratum* seedlings in aeroponic culture to treatments of different ammonium: Nitrate ratios. *Journal of Plant Nutrition* 40(20): 2850–2861.

Zsoldos, F. et al., 1987, Lack of active K+ uptake in aeroponically grown wheat seedlings. *Physiologia Plantarum* 71: 6, 359–364.

5 Aeroponics Innovations

Every once in a while, a new technology, an old problem, and a big idea turn into an innovation.

Dean Kamen

The pioneers in aeroponic technology have been Disneyworld in EPCOT and NASA. These organizations have been the innovators and have launched this technology in last 30 years.

AEROPONICS AT DISNEY WORLD

In the 1990s, Walt Disney World launched the Land as part of the EPCOT park. It became the place where innovative farming technology was developed. There were 2.5 million square feet of active greenhouses which produced enormous amounts of the food served in the park's restaurants while providing a space for scientists to research indoor plant growing techniques. Among those techniques are vertical growing techniques. They asked the key questions—is there a way to increase food production while using less water, less fertilizer, fewer pesticides, and even less space. Plants are grown vertically using either stacked gardens or specialized trellises that allow crops to reach gravity-defying heights. Produce grown in this way uses a fraction of the space required by conventional methods, which results in saving water and increasing yields. Another groundbreaking way to grow food more efficiently is to grow it without soil. Hydroponics is a method of growing plants using just water and nutrients, and Disney uses this technology throughout their greenhouses (Lynch, 2018).

Epcot produces over 27,000 heads of lettuce a year using a form of hydroponics called the nutrient film technique. This technique allows a shallow stream of nutrient-infused water to circulate past the plants' exposed roots, providing all the good stuff a plant would usually get from the soil. However, it is quicker than growing crops in the soil, it uses less water, and it allows growers to stack plants closer together. Aeroponics is another form of dirt-free farming at Epcot. It is similar to hydroponics, but instead of feeding the plants with a stream of water, their roots dangle freely in the air and are periodically sprayed with a fine mist of atomized nutrients. Epcot has partnered with NASA to pioneer new ways of growing food aeroponically because it is exactly the sort of technology that could be used for a long-term space mission.

AEROPONICS AT NASA

Plants have been to space since 1960, but NASA's plant growth experiments began in earnest during the 1990s (Spinoff, 2007). Experiments aboard the space shuttle and International Space Station have exposed plants to the effects of microgravity. These experiments use the principles of aeroponics: growing plants in an air/mist environment with no soil and very little water (NASA).

In 1997, NASA-sponsored studies aboard the Mir space station using Adzuki bean seeds and seedlings, a high-protein Asian food crop. While the beans were growing in zero gravity, ground control experiments were performed to see how another group of seeds and seedlings responded on Earth. Both sets of plants were treated with an all-natural, organically derived, disease control liquid known as Organic Disease Control or Organically Derived Colloidals (ODC).

While all of the seeds grew well, those aboard Mir grew more than those on Earth. Both sets of plants treated with the ODC method grew more robustly and exhibited less fungal infection than the untreated seeds and seedlings.

Results from NASA's research aboard Mir have contributed to rapid-growth systems now used on Earth. Plants are sown from either cuttings or seeds and then suspended mid-air in a growing chamber. The developing root systems grow under an enclosed, air-based environment, which is regularly misted with a fine, nutrient-rich spray. Aeroponic growing systems provide clean, efficient, and rapid food production. Crops can be planted and harvested in the system year round without interruption and without contamination from soil, pesticides, and residues. Since the growing environment is clean and sterile, it greatly reduces the chances of spreading plant disease and infection commonly found in soil and other growing media.

The suspended system also has other advantages. Seedlings do not stretch or wilt while their roots are forming. Once the roots are developed, the plants can be easily moved into any type of growing media without the risk of transplant shock, which often sets back normal growth.

Aeroponics systems can reduce water usage by 98%, fertilizer usage by 60%, and pesticide usage by 100%, all while maximizing crop yields. Plants grown in the aeroponic systems have also been shown to uptake more minerals and vitamins, making the plants healthier and potentially more nutritious.

Tomato growers traditionally sow their plants in pots, waiting at least 28 days before transplanting them into the ground. Using an aeroponic system, growers can sow the plants in the growing chamber, and then transplant them just 10 days later. This advanced technology produces six tomato crop cycles per year, instead of the traditional one to two crop cycles.

Successful long-term missions into deep space will require that crews grow some of their own food during flight. Aeroponic crops are also a potential source of fresh oxygen and clean drinking water. But this is about more than a breath of fresh air or taking a quick shower. Each ounce of food and water produced aboard a spacecraft reduces payload weight, allowing space for other cargo that cannot be produced onboard.

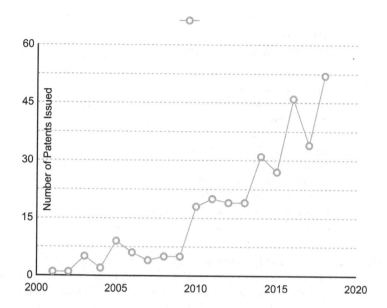

FIGURE 5.1 Number of US Patents related to aeroponics issued from 2001 to 2018.

There are several ways to measure innovation. The pioneers, Disneyworld and NASA, have been summarized and next the intellectual property related to aeroponics will be reviewed with a focus on the US Patents issued since 2001. Figure 5.1 shows the number of patents related to aeroponics by the year the patent was issued. The graph follows a similar trend to the peer-reviewed articles published since 1970 (Figure 4.1) in Chapter 4 on the Science of Aeroponics. The number of patents seems to be increasing by year which may mean that this technology is increasing in value and applicability as a sustainable commercial agricultural method.

Of the 308 patents issued since 2001–2018, 44% were assigned to a company, research foundation, or university. The remainder are held by private individuals. The top ten assignees are listed in Table 5.1. The leader is Xyleco with 23 patents. Xyleco was founded by Marshall Medoff with the vision to increase the worlds sustainable resources. It is a privately held research and manufacturing company started in Woburn, MA and moved to Wakefield in 2015. Xyleco is developing a process to convert biomass into useful products, including cellulosic ethanol. Xyleco's global intellectual property portfolio exceeds 5,000 patents and applications in 100 countries worldwide (Xyleco website).

Xyleco's Board of Directors includes three Nobel prize winners in science (chemistry and physics), a senior professor in chemical engineering at MIT, a former US Secretary of Energy and a former Chief Executive Officer of the world's largest oil company. Their patents describe methods that can be used to process biomass using hydroponics. Hydroponics is a method of growing plants using mineral nutrient solutions, without soil. Plants may be grown with their roots in the

TABLE 5.1

Top 10 Aeroponic Patent Holders

Xyleco, Inc.	23
Bayer Crop Science AG	17
Aerogrow International, Inc.	12
Indigo AG, Inc.	9
Living Greens Farm, Inc.	7
Bayer Intellectual Property GMBH	5
Cal Safe Soil, LLC	5
Colorado Energy Research Technologies, LLC	5
Research Development Foundation	5
Indoor Farms of America, LLC	4

mineral nutrient solution only (solution culture) or in an inert medium (medium culture) such as perlite, gravel, or mineral wool. The three main types of solution cultures are static solution culture, continuous flow solution culture, and aeroponics.

Xyleco is followed by Bayer CropScience with 17 and Bayer Intellectual Property GMBH with five additional patents (CropScience website). So Bayer holds 22 patents. Patents describe methods and compositions for the control of rust fungi by inhibiting specific genes. Applications may be made as seed treatment, foliar application, stem application, drench or drip application (chemigation) to the seed, the plant or the fruit, to the soil or to inert substrate.

AeroGrow holds 12 patents which are focused on home use aeroponic devices.

A countertop gardening appliance dimensioned to fit on a countertop and a controller adapted to activate the artificial light source and the gas pump on predetermined time cycles. AeroGrow International, Inc. is the manufacturer and distributor of an indoor gardening system—the AeroGarden line of Smart Countertop Gardens. Headquartered in Boulder, Colorado, AeroGrow has a commitment to helping people successfully grow healthy, fresh food regardless of season, location, or experience (Aerogrow website).

Since introducing the first AeroGarden in March 2006, AeroGrow has expanded the product line to include multiple gardens with different price points and consumer benefits. From small to tall and from simple to sophisticated, AeroGarden provides products for every consumer regardless of their aptitude or gardening experience.

AeroGrow also develops, manufactures, and markets a variety of consumable products for use in its gardens including Seed Kits, Grow Lights, hydroponic nutrients, and accessory products. On April 23, 2013, the Company announced that Scotts Miracle-Gro made a $4.5 million equity investment and IP acquisition with the Company, resulting in a 30% beneficial ownership interest in AeroGrow. In the process, AeroGrow took steps to entirely eliminate its long-term debt, restructured the balance sheet to facilitate potential future transactions, and gained a valuable partnership for growth. The agreement affords AeroGrow the use of the globally recognized and highly trusted Miracle-Gro brand name while also providing AeroGrow a broad base of support in marketing, distribution, supply chain logistics, R&D, and sourcing.

AeroGrow was founded in July 2002 and became a publicly-traded company on February 24, 2006. AeroGrow is headquartered in Boulder, Colorado and employs approximately 25 people.

Indigo AG holds nine patents mainly focused on the microbiome but also has a patent on a modular aeroponic system. The history of the company starts with Flagship Venture Labs, the innovation foundry of Flagship Ventures, which began research in 2012 and created Symbiota in 2014. Symbiota then became Indigo. They sought to recreate the plant's natural microbial makeup, involving the identification and sequence 40,000 endosymbionts—microbes that live inside the plant itself—creating the largest body of data on earth for these microbes. To reintroduce these beneficial microbes back into crops, they created a seed coating that provides a path for these microbes to return to their native habitat. This in turn yields more abundant, healthier crops that are more resistant to stresses like insufficient water, low nitrogen, high temperature, and salty soils, while bolstering crops' resistance to disease and harmful insects (6).

Most of their patents focus on the endophyte populations are intended to be useful in the improvement of agricultural plants, and as such, may be formulated with other compositions as part of an agriculturally compatible carrier. It is contemplated that such carriers can include, but not be limited to: seed treatment, root wash, seedling soak, foliar application, soil inocula, in-furrow application, sidedress application, soil pretreatment, wound inoculation, drip tape irrigation, vector-mediation via a pollinator, injection, osmopriming, hydroponics, aquaponics, and aeroponics. The carrier composition with the endophyte populations, may be prepared for agricultural application as a liquid, a solid, or a gas formulation. Application to the plant may be achieved, for example, as a powder for surface deposition onto plant leaves, as a spray to the whole plant or selected plant element, as part of a drip to the soil or the roots.

One patent describes a modular aeroponic system that accommodates different support-media and misting or spray configurations. It comprises a root chamber with plumbing that is coupled to a nutrient distribution system and a misting system.

Living Greens Farms Inc. hold seven aeroponic-related patents. It consists of a vertical triangular shaped aeroponic system and irrigation system. It consists of a chamber into which plant roots may extend, and into which one or more exit ports are able to grow their plants (Living Greens website).

Faribault, Minnesota-based Living Greens Farm Inc., which runs one of the largest indoor farms in the world, is looking to turn a new leaf. The produce grower has so far raised $3 million in new capital as part of a $12 million Series A round. Once fundraising is complete, Living Greens Farm CEO Dana Anderson will use the money to grow the scale of its operations.

The farm currently has 20,000 square feet of grow space, allowing for annual production of about 760,000 heads of lettuce. The $12 million infusion would help the company expand and re-design their space for more efficiency. Ultimately, Anderson said he would like to see the space grow to about 60,000 square feet of farm space—a possibility in their 45,000 square-foot building because the vertical structure of the farm magnifies floor space. Living Greens Farm is not your typical farm. Rather, it grows plants indoors using aeroponics, which involves the practice of suspending a plant's roots in the air and spraying them with a nutrient-rich solution instead of burying them in soil. The method is said to be the fastest way to grow greens. The company also uses a computer system to control elements such as light, temperature, humidity, and carbon dioxide, and has its own patented vertical growing and traversing misting systems. Living Greens Farm claims its system uses 200 times less land and 95% less water than a traditional farm. The enhanced growing area that the Series A capital could multiply Living Greens Farm production to 3 million heads of lettuce annually. Anderson noted that production level would be equal to that of a roughly 200-acre farm.

Furthermore, Anderson said the $12 million will be used to commercialize Living Greens Farm's technology. The company currently holds four patents and though Anderson said they'll continue to invent new assets, they are focused primarily on refining their existing technology and proliferating it across the world. "There continues to be a lot of issues with field-grown products," he said, citing instances of recalls of romaine and people getting sick from bad lettuce. "We're offering the safer, healthier option consumers are looking for."

The companies with the next highest number of patents in this field are:

Cal Safe Soil LLC (focused on nutrient compositions)
Colorado Energy Research Corp (fermentation methods)
Research Development Corp (saponin mixtures and compounds from Acacia that exhibit potent anti-tumor effects against a variety of tumor cells)
Indoor Farms of America (vertical aeroponic systems)

Other companies that have patents on aeroponic growing systems in alphabetic order:

Aero Development Corp USA (commercial vertical aeroponic systems)
Aessense Technology USA (sensors for aeroponic systems)
Agricool France (Container system)
AirGrow IP USA (Apparatus for aeroponically growing and development plants)
AliGroWorks USA, Inc. (automated watering system and plant growing system)
Azoth Solutions, LLC USA (vertical aeroponic cultivation system)
Croptimize APS Denmark (hydroponic/aeroponic horizontal stacked vertically system)
Double J Holdings, LLC USA (hydroponic/aeroponic horizontal stacked vertically system)
Econow Systems, LLC USA (self-contained compact automated aeroponic systems)

EZ-Clone Enterprises, Inc. USA (aeroponic horizontal stacked vertically system)

GnomeWorks Greenhouse Inc. Canada (aeroponic horizontal stacked vertically grow tube system)

Green Thumb Technology Inc. USA (vertical hydroponic/aeroponic system)

GrowX Inc. USA (vertical aeroponic system)

Harvest Urban Farms Inc. USA (commercial aeroponic horizontal stacked vertically system)

Intelligent Growth Solutions Ltd. UK (aeroponic horizontal stacked vertically system)

Inventagon LLC USA (horizontal fabric based aeroponic system)

Land Green and Technology Co. Ltd. Taiwan (aeroponic horizontal stacked vertically system)

Living Box Ltd. USA (horizontal hydroponic/aeroponic system)

Microwave Ltd. Gibraltar (horizontal hydroponic/aeroponic system)

MJNN LLC USA (vertical hydroponic/aeroponic curved wall system)

Ohneka Farms LLC USA (vertical hydroponic/aeroponic system)

Opusculum PTE Ltd. Singapore (vertical aeroponic system using perforated materials)

Orbital Technologies, Inc. USA (horizontal rotary cylinder hydroponic/aeroponic system)

Perfect Plant LLC Cook Island (hydroponic/aeroponic system)

Phidro LLC USA (vertically angled modular areohydroponic system)

Plantui Oy USA (indoor aerohydroponic horizontal system)

RoBotany Ltd. USA (vertically stacked aerohydroponic horizontal system)

SproutslO, Inc. USA (aerohydroponic system)

STMicroelectronics, Inc. USA (single plant misted aeroponic system)

Sustainable Strategies LLC West Indies (overhead spray aerohydroponic system)

University of Guelph Canada (vegetated mats in an aerohydroponic horizontal system)

Xiamen SUPERPRO Technology CO., Ltd. China (hydroponic system)

Other companies have patents related to aeroponics but are focused on nutrient solution compositions, fungicides, sensors, artificial lights, materials, pesticides, greenhouses, and other concomitant topics.

In addition to the company assigned patents there are 56% of the patents that were not assigned to any company. Of those 17% were plant growing systems and the word 'aeroponic' was used in the patent title in 6% of those. Many of these patents were hydroponic systems or variations on hydroponic systems and the word 'aeroponic' was used in the patent description, but in most cases was not the focus of the patent.

Table 5.2 is a list of the patents by number, inventor, assignee, and date of issue. These are sorted in chronological order from the most current to the oldest based on date of issue.

TABLE 5.2

US Patents Related to Aeroponics Since 2001 by Number, Title, Principal Investigator, and Date of Issue

Patent Number	Title	Principal Investigator	Date of Issue
20190021247	Aqueous Grow Chamber Recirculating Nutrient Control System and Sensor Calibration	Boerema; Martin; et al.	January 24, 2019
20190008158	Endophyte Compositions and Methods for Improvement of Plant Traits in Plants of Agronomic Importance	On Maltzahn; Geoffrey	January 10, 2019
20190008104	Root Trellis for Use in Hydroponic Growing and Methods of Using Same	Sabzerou; Nate; et al.	January 10, 2019
20180368346	Fogponic Plant Growth System	Watson; Michael C.	December 27, 2018
20180368343	Sustainable Growing System and Method	O'Rourke; Greg	December 27, 2018
20180361420	Low Frequency Electrostatic Ultrasonic Atomizing Nozzle	Gao; Jianmin; et al.	December 20, 2018
20180354868	Soluble Fertilizer Formulation and Method for Use Thereof	Taganov; Igor; et al.	December 13, 2018
20180352761	Bio Cell System	Hall; David R.; et al.	December 13, 2018
20180347406	Quintuple-Effect Generation Multi-Cycle Hybrid Renewable Energy System With Integrated Energy Provisioning, Storage Facilities And Amalgamated Control System	Friesth; Kevin Lee	December 6, 2018
20180338440	Cloner	Dearinger; Robert	November 29, 2018
20180338439	Temperature and Light Insulated Aeroponics Root Chamber Built with Opaque High-Density Expanded Polypropylene	Gao; Wanjun	November 29, 2018
20180338430	Aeroponic Irrigation System	Cobzev; Anton N.; et al.	November 29, 2018
20180332777	Vegetation Grow Light Embodying Power Delivery and Data Communication Features	Thosteson; Eric; et al.	November 22, 2018
20180325056	Aeroponic Growing Column and System	Stolzfus; Samuel A.; et al.	November 15, 2018
20180325055	Aeroponic Apparatus	Krakover; Gilad	November 15, 2018

(Continued)

TABLE 5.2 (*Continued*)

US Patents Related to Aeroponics Since 2001 by Number, Title, Principal Investigator, and Date of Issue

Patent Number	Title	Principal Investigator	Date of Issue
20180325052	Aeroponic Installation, System and Freight Container	GRU; Gonzague; et al.	November 15, 2018
20180325038	Automated Vertical Plant Cultivation System	Spiro; Daniel S.	November 15, 2018
20180317411	Automated Vertical Plant Cultivation System	Spiro; Daniel S.	November 8, 2018
20180317410	Automated Vertical Plant Cultivation System	Spiro; Daniel S.	November 8, 2018
20180317409	Vertical Aeroponic Growing Apparatus	Staffeldt; Benjamin Jon	November 8, 2018
20180308028	Control of Plant-Growing Facilities And Workforces	Zhang; Hong; et al.	October 25, 2018
20180298398	Methods and Compositions for the Control of Rust Fungi by Inhibiting Expression of the HXT1 Gene	Gautier; Pierrick; et al.	October 18, 2018
20180295800	Vertically Oriented Modular Aerohydroponic Systems and Methods of Planting and Horticulture	Kiernan; John Thomas	October 18, 2018
20180288954	Method of Growing Plants and System Therefore	Glaser; Grant; et al.	October 11, 2018
20180288949	Method of Growing Plants, Growing Chamber and System Therefore	Glaser; Grant; et al.	October 11, 2018
20180251776	Endophyte Compositions and Methods for Improvement of Plant Traits	Riley; Raymond	September 6, 2018
20180249716	Endophyte Compositions and the Methods for Improvement of Plant Traits	Riley; Raymond	September 6, 2018
20180242539	An Intelligent Integrated Plant Growth System and a Process of Growing Plant Thereof	Bhattacharya; Deb Ranjan; et al.	August 30, 2018
20180242531	Device for Promoting Root Function in Industrial Farming	Berry, III; Henry Kistler; et al.	August 30, 2018
20180237352	Soluble Fertilizer Formulation and Method for Use Thereof	Taganov; Igor; et al.	August 23, 2018
20180220603	Industrial Aeroponics	Burford; Gerek Levi	August 9, 2018
20180213800	Penicillium Endophyte Compositions and Methods for Improved Agronomic Traits in Plants	Djonovic; Slavica; et al.	August 2, 2018
20180206420	Irrigation Device	Kendall; John William	July 26, 2018

(Continued)

TABLE 5.2 (*Continued*)
US Patents Related to Aeroponics Since 2001 by Number, Title, Principal Investigator, and Date of Issue

Patent Number	Title	Principal Investigator	Date of Issue
20180177196	Fungal Endophytes for Improved Crop Yields and Protection from Pests	Sword; Gregory A.	June 28, 2018
20180177143	Soil-less Pre-Vegetated Mat and Process for Production Thereof	Yuristy; Greg; et al.	June 28, 2018
20180153174	Modulated Nutritional Quality Traits in Seeds	Riley; Raymond; et al.	June 7, 2018
20180153115	Apparatus for Crop/Plant/Life-Form Cultivation	Edke; Rajesh; et al.	June 7, 2018
20180153114	Modular Aeroponic System and Related Methods	Riley; Raymond; et al.	June 7, 2018
20180132434	Method and System for Capable of Selecting Optimal Plant Cultivation Method	Fu; Lid	May 17, 2018
20180105827	Method for Producing High-Quality Recombinant Allergens in a Plant	Gomord; Veronique; et al.	April 19, 2018
20180098515	Controlled Environment and Method	Anderson; Dana; et al.	April 12, 2018
20180092314	Vertical Hydroponics Systems	McGuinness; Jennifer; et al.	April 5, 2018
20180084744	Column Element for a Device for the Vertical Cultivation of Plants	Tidona; Marco	April 12, 2018
20180077884	Apparatus and Method for Automated Aeroponic Systems For Growing Plants	Barker; Stephen F.; et al.	March 22, 2018
20180070545	Ventusponic Automated Hydroponic Plant Growth Apparatus and Method	Perry; Stanley J.; et al.	March 15, 2018
20180064048	Growing Systems and Methods	Joseph; Timothy E.; et al.	March 8, 2018
20180042191	Aeroponics System with Rack and Tray	Blackburn; William J.; et al.	February 15, 2018
20180020677	Seed Endophytes across Cultivars and Species, Associated Compositions, and Methods of Use Thereof	Ambrose; Karen V.; et al.	January 25, 2018
20180014486	Control and Sensor Systems for an Environmentally Controlled Vertical Farming System	Creechley; Jaremy; et al.	January 18, 2018
20180014485	Environmentally Controlled Vertical Farming System	Whitcher; John L.; et al.	January 18, 2018
20180014471	Vertical Growth Tower and Module for an Environmentally Controlled Vertical Farming System	Jensen; Rob; et al.	January 18, 2018

(*Continued*)

TABLE 5.2 (*Continued*)

US Patents Related to Aeroponics Since 2001 by Number, Title, Principal Investigator, and Date of Issue

Patent Number	Title	Principal Investigator	Date of Issue
20180007849	Hydroculture System	Cohen; Mordehay Shlomo; et al.	January 11, 2018
20180007845	Systems, Methods, and Devices for Aeroponic Plant Growth	Martin; John-Paul Armand	January 11, 2018
20180000029	Vertical Aeroponic Plant Growing Enclosure	Martin; David W.; et al.	January 4, 2018
20170347548	System and Method for Cultivating Plants	Jollie; Jesse L.	December 7, 2017
20170339846	Apparatus and Method for Autonomous Controlled Environment Agriculture	Lawrence; Austin Blake; et al.	November 30, 2017
20170332567	Aerobic, Bioremediation Treatment System Comprising Floating Inert Media in an Aqueous Environment	Gencer; Mehmet A.; et al.	November 23, 2017
20170332544	Data Driven Indoor Farming Optimization	Conrad; Travis Anthony; et al.	November 23, 2017
20170321233	Processing Biomass	Medoff; Marshall	November 9, 2017
20170208757	Horticultural Nutrient Control System and Method for Using Same	Valmont; Justin Jean Leonard	July 27, 2017
20170202163	Aeroponic Culture Line	Aschheim; Raymond; et al.	July 20, 2017
20170188531	Accelerated Plant Growth System	Daniels; John J.	July 6, 2017
20170188526	Aeroponics System with Microfluidic Die and Sensors for Feedback Control	De Fazio; Marco; et al.	July 6, 2017
20170181394	Hydroponic Growing System	Kotsatos; Vasilios M.	June 29, 2017
20170152532	Processing Biomass	Medoff; Marshall	June 1, 2017
20170152468	Biodigestion Reactor, System Including the Reactor, and Methods of Using Same	Fernandez; Todd; et al.	June 1, 2017
20170150687	Device for Hydroponic Cultivation	Loiske; Janne; et al.	June 1, 2017
20170150686	Container for Supplying Plant Roots with Nutrient Solution without the Use of Soil	Erbacher; Clemens	June 1, 2017
20170150684	Hydroponic Indoor Gardening Method	Vuorinen; Kari; et al.	June 1, 2017

(*Continued*)

TABLE 5.2 (*Continued*)
US Patents Related to Aeroponics Since 2001 by Number, Title, Principal Investigator, and Date of Issue

Patent Number	Title	Principal Investigator	Date of Issue
20170137311	System and Method of Treating Organic Material	Fernandez; Todd; et al.	May 18, 2017
20170130252	Processing Biomass	Medoff; Marshall	May 11, 2017
20170105373	Fluid Filtration and Distribution System for Planting Devices	Byron, III; James Edgar; et al.	April 20, 2017
20170105368	Hybrid Hydroponic Plant Growing Systems	Mehrman; Edward L.	April 20, 2017
20170105358	Led Light Timing in a High Growth, High Density, Closed Environment System	Wilson; James G.	April 20, 2017
20170099791	Growing Systems and Methods	Joseph; Timothy E.; et al.	April 13, 2017
20170099790	Modular Automated Growing System	Gonyer; Daegan; et al.	April 13, 2017
20170097329	Optical Sensor for Fluid Analysis	Smeeton; Tim Michael; et al.	April 6, 2017
20170094920	Integrated Incubation, Cultivation and Curing System and Controls for Optimizing and Enhancing Plant Growth, Development and Performance of Plant-Based Medical Therapies	Ellins; Craig; et al.	April 6, 2017
20170088477	Nutrient Rich Compositions	Morash; Daniel M.; et al.	March 30, 2017
20170086399	Irrigation System	Anderson; Dana; et al.	March 30, 2017
20170064912	Growing Vegetables within a Closed Agricultural Environment	Tabakman; Zale Lewis David	March 9, 2017
20170055472	Rotary Plant Growing Apparatus	Gunther; Matt	March 2, 2017
20170055461	Automated Hydroponics System and Method	Neuhoff J. R.; Robert V.; et al.	March 2, 2017
20170045488	Methods and Apparatus for Determining Fertilizer/Treatment Requirements and/or Predicting Plant Growth Response	Riess; Mark James; et al.	February 16, 2017
20170035008	Method for Optimizing and Enhancing Plant Growth, Development and Performance	Ellins; Craig; et al.	February 9, 2017

(*Continued*)

TABLE 5.2 (*Continued*)

US Patents Related to Aeroponics Since 2001 by Number, Title, Principal Investigator, and Date of Issue

Patent Number	Title	Principal Investigator	Date of Issue
20170035002	Apparatus for Optimizing and Enhancing Plant Growth, Development and Performance	Ellins; Craig; et al.	February 9, 2017
20170023193	Grow Light Embodying Power Delivery and Data Communications Features	Thosteson; Eric; et al.	January 26, 2017
20170013810	Portable Agrarian Biosystem	Grabell; Herb; et al.	January 19, 2017
20160366892	Designed Complex Endophyte Compositions and Methods for Improved Plant Traits	Ambrose; Karen V.; et al.	December 22, 2016
20160324086	Aeroponic Growth Unit for Growing Plants, System, Greenhouse and Methods Thereof	Steuart; Douglas Osborne	November 10, 2016
20160318820	Plant Nutrient Coated Nanoparticles and Methods for their Preparation and Use	Deb; Nilanjan	November 3, 2016
20160316760	Isolated Complex Endophyte Compositions and Methods for Improved Plant Traits	Ambrose; Karen V.; et al.	November 3, 2016
20160316645	Hydroponic System	Neufeld; Jon	November 3, 2016
20160298265	Spider Silk and Synthetic Polymer Fiber Blends	Lewis; Randolph V.; et al.	October 13, 2016
20160298147	Processing Biomass	Medoff; Marshall	October 13, 2016
20160295820	Automatic Tower with Many Novel Applications	Aykroyd; Henry; et al.	October 13, 2016
20160289710	Processing Biomass	Medoff; Marshall	October 6, 2016
20160289709	Processing Biomass	Medoff; Marshall	October 6, 2016
20160270311	Vertical Aeroponic Plant Growing Enclosure with Support Structure	Martin; David W.; et al.	September 22, 2016
20160262391	Pesticidal Composition Comprising a Pyridylethylbenzamide Derivative and an Insecticide Compound	Hungenberg; Heike; et al.	September 15, 2016
20160255781	Wireless Sensor Systems for Hydroponics	Chen; Tianshu; et al.	September 8, 2016
20160235025	Aeroponic Cultivation System	Bray; Jon Aaron	August 18, 2016

(*Continued*)

TABLE 5.2 (*Continued*)

US Patents Related to Aeroponics Since 2001 by Number, Title, Principal Investigator, and Date of Issue

Patent Number	Title	Principal Investigator	Date of Issue
20160235024	Flexible Hydroponics Growing Model and System	Xu; Liqiang; et al.	August 18, 2016
20160227719	Aeroponic System	Orff; Dylan	August 11, 2016
20160215251	Systems and Methods of Improved Fermentation	Powell; Wayne J.; et al.	July 28, 2016
20160214945	Agricultural Chemicals	Urch; Christopher; et al.	July 28, 2016
20160212997	Agricultural Chemicals	Urch; Christopher; et al.	July 28, 2016
20160198650	Compact Plant Growing Unit	Lahaeye; Eric	July 14, 2016
20160192594	Central Processing Horticulture	Mawendra; Aravinda Raama	July 7, 2016
20160183487	System and Method for Detecting the Absence of a Thin Nutrient Film in a Plant Growing Trough	Kabakov; Bentsion	June 30, 2016
20160183486	System and Method for Growing Crops and Components Therefor	Kabakov; Bentsion	June 30, 2016
20160183476	Mobile Plant Service System and Components Therefor	Kabakov; Bentsion	June 30, 2016
20160174546	Nano Particulate Delivery System	Berg; Paulo Sergio; et al.	June 23, 2016
20160165926	Processing Biomass	Medoff; Marshall	June 16, 2016
20160165925	Nutrient Rich Compositions	Morash; Daniel M.; et al.	June 16, 2016
20160159705	Nutrient Rich Compositions	Morash; Daniel M.; et al.	June 9, 2016
20160135395	Modular Hydroponic Growing System	Umpstead; Neil	May 19, 2016
20160135394	Hydroponic System For Growing Plants	Wagner; Daniel Davidson	May 19, 2016
20160128289	Hydroponic System with Actuated Above-Plant Platform	Wong; Simon; et al.	May 12, 2016
20160128288	Self-Watering, Self-Lighting Hydroponic System	Pettinelli; Gabrielle; et al.	May 12, 2016
20160120141	Aeroponic Growing Column and System	Stolzfus; Samuel A.; et al.	May 5, 2016

(*Continued*)

TABLE 5.2 (*Continued*)

US Patents Related to Aeroponics Since 2001 by Number, Title, Principal Investigator, and Date of Issue

Patent Number	Title	Principal Investigator	Date of Issue
20160088810	Apparatus for Growing Plants Hydroponically	Rabii; Mehdi; et al.	March 31, 2016
20160083309	System, Method, and Composition for Enhancing Solutions from Bioreactors for Processes Including Liquid Fertilizer Preparation and Nutrient Extraction	Pina; Tinia	March 24, 2016
20160081281	Active Polymer Material for Agriculture Use	Horinek; David	March 24, 2016
20160054281	Optical Sensor for Fluid Analysis	Smeeton; Tim Michael; et al.	February 25, 2016
20160029581	Vertical Aeroponic Plant Growing Enclosure with Support Structure	Martin; David W.; et al.	February 4, 2016
20160029578	Aeroponic Commercial Plant Cultivation System Utilizing a Grow Enclosure	Martin; David W.; et al.	February 4, 2016
20160021837	Apparatus for Providing Water and Optionally Nutrients to Roots of a Plant And Method of Using	Kernahan; Kent	January 28, 2016
20160021836	Aeroponic Growth System Wireless Control System and Methods of Using	Kernahan; Kent	January 28, 2016
20160017225	Nutrient Rich Compositions	Morash; Daniel M.; et al.	January 21, 2016
20160016944	Fungicidal 3—Heterocycle Derivatives	Rebstock; Anne-Sophie; et al.	January 21, 2016
20160007607	Fungicidal 3—Oxadiazolone Derivatives	Rebstock; Anne-Sophie; et al.	January 14, 2016
20160002217	Fungicidal 3—Oxadiazolone Derivatives	Rebstock; Anne-Sophie; et al.	January 7, 2016
20160000078	Fungicidal 3—Oxadiazolone Derivatives	Rebstock; Anne-Sophie; et al.	January 7, 2016
20150361143	Method for Producing High-Quality Recombinant Allergens in a Plant	Gomord; Veronique; et al.	December 17, 2015
20150359181	Soil-less Pre-Vegetated Mat and Process for Production Thereof	Yuristy; Greg; et al.	December 17, 2015
20150353974	Processing Biomass	Medoff; Marshall	December 10, 2015
20150342224	Processing Biomass	Medoff; Marshall	December 3, 2015

(*Continued*)

TABLE 5.2 (*Continued*)

US Patents Related to Aeroponics Since 2001 by Number, Title, Principal Investigator, and Date of Issue

Patent Number	Title	Principal Investigator	Date of Issue
20150334930	Modular Aeroponic Growing Column and System	Stoltzfus; Samuel A.; et al.	November 26, 2015
20150334929	Self-Contained Aeroponic Growing Unit	Miller; Scott	November 26, 2015
20150320047	Pesticide Composition Comprising Fosetyl-Aluminium, Propamocarb-Hcl and an Insecticide Active Substance	Hungenberg; Heike; et al.	November 12, 2015
20150315466	Nutrient Rich Compositions	Morash; Daniel M.; et al.	November 5, 2015
20150305313	Integrated Multi-Trophic Farming Process	Licamele; Jason	October 29, 2015
20150305258	Methods and Apparatus for a Hybrid Distributed Hydroculture System	Broutin Farah; Jennifer; et al.	October 29, 2015
20150292120	Apparatus and Methods for Producing Fibers from Proteins	Lewis; Randolph V.; et al.	October 15, 2015
20150257387	Pesticide Composition Comprising a Tetrazolyloxime Derivative and a Fungicide or an Insecticide Active Substance	Coqueron; Pierre-Yves; et al.	September 17, 2015
20150237807	Continuous Bioprocess for Organic Greenhouse Agriculture	Valiquette; Marc-Andre	August 27, 2015
20150223418	Light-Weight Modular Adjustable Vertical Hydroponic Growing System and Method	Collins; Fred; et al.	August 13, 2015
20150196002	Automated Hybrid Aquaponics and Bioreactor System Including Product Processing and Storage Facilities with Integrated Robotics, Control System, and Renewable Energy System Cross-Reference to Related Applications	Friesth; Kevin	July 16, 2015
20150173315	Method and Modular Structure for Continuously Growing an Aeroponic Crop	Aznar Vidal; Carlos	June 25, 2015
20150143806	Quintuple-Effect Generation Multi-Cycle Hybrid Renewable Energy System with Integrated Energy Provisioning, Storage Facilities and Amalgamated Control System Cross-Reference to Related Applications	Friesth; Kevin Lee	May 28, 2015

(Continued)

TABLE 5.2 (*Continued*)

US Patents Related to Aeroponics Since 2001 by Number, Title, Principal Investigator, and Date of Issue

Patent Number	Title	Principal Investigator	Date of Issue
20150102137	Irrigation System	Anderson; Dana; et al.	April 16, 2015
20150101082	Processing Seeds by Cold Plasma Treatment to Reduce an Apparent Contact Angle of Seeds Coat Surface	Bormashenko; Edward; et al.	April 9, 2015
20150082495	RNAi for the Control of Fungi and Oomycetes by Inhibiting Saccharopine Dehydrogenase Gene	Delebarre; Thomas; et al.	March 19, 2015
20150068122	Commercial Aeroponics System	Juncal; Shaun; et al.	March 12, 2015
20150045221	Composition Comprising an Elicitor of the Plant Immune System	Buonatesta; Raffaele; et al.	February 12, 2015
20150040477	Aquaponic Growth Bucket	Wang; Shumin; et al.	February 12, 2015
20150031730	Fungicidal 3-[(1,3-thiazol-4-ylmethoxyimino)(phenyl)methyl]-2-substituted–1,2,4-oxadiazoL-5(2H)-one Derivatives	Braun; Christoph;; et al.	January 29, 2015
20150027051	Controlled Environment and Method	Anderson; Dana; et al.	January 29, 2015
20150008191	Low-Turbulent Aerator and Aeration Method	Ladouceur; Richard	January 8, 2015
20150000190	Automated Plant Growing System	Gibbons; Adam	January 1, 2015
20140350058	Fungicidal 3-[(pyridin-2-ylmethoxyimino)(phenyl)methyl]-2-substituted-1,2,-4-oxadiazol-5(2h)-one Derivatives	Braun; Christoph;; et al.	November 27, 2014
20140349851	Plant Growth Enhancing Mixture and Method of Applying Same	Stoller; Jerry; et al.	November 27, 2014
20140349848	Fungicidal 4-substituted-3–1,2,4-oxadizol-5(4h)-one Derivatives	Braun; Christoph; et al.	November 27, 2014
20140348982	Processing Biomass	Medoff; Marshall	November 27, 2014
20140342908	Plant Growth Enhancing Mixture and Method of Applying Same	Stoller; Jerry; et al.	November 20, 2014
20140338261	Modular aeroponic system and related methods	Sykes; Chad Colin	November 20, 2014
20140325910	Traveling Seed Amplifier, TSA, Continuous Flow Farming of Material Products, MP	Faris; Sadeg M.	November 6, 2014

(*Continued*)

TABLE 5.2 (*Continued*)
US Patents Related to Aeroponics Since 2001 by Number, Title, Principal Investigator, and Date of Issue

Patent Number	Title	Principal Investigator	Date of Issue
20140325909	Permeable Three Dimensional Multi-Layer Farming	Faris; Sadeg M.	November 6, 2014
20140325908	High Density Three Dimensional Multi-Layer Farming	Faris; Sadeg M.	November 6, 2014
20140311030	Irrigation System	Anderson; Dana; et al.	October 23, 2014
20140311029	Controlled Environment and Method	Anderson; Dana; et al.	October 23, 2014
20140287921	Plant Growth Enhancing Mixture and Method of Applying Same	Stoller; Jerry; et al.	September 25, 2014
20140287919	Modulation of Plant Immune System Function	Levine; Robert B.; et al.	September 25, 2014
20140259920	LED Light Timing in a High Growth, High Density, Closed Environment System	Wilson; James G.	September 18, 2014
20140237897	Means and Methods for Growing Plants in High Salinity or Brackish Water	Lotvak; Izhak Levi; et al.	August 28, 2014
20140235442	Pesticidal Composition Comprising a Pyridylethylbenzamide Derivative and an Insecticide Compound	Hungenberg; Heike; et al.	August 21, 2014
20140221210	Acyl-Homoserine Lactone Derivatives for Improving Plant Yield	Dahmen; Peter; et al.	August 7, 2014
20140206726	Fungicide Hydroximoyl-Tetrazole Derivatives	Benting; Juergen; et al.	July 24, 2014
20140178980	Systems and Methods of Improved Fermentation	Powell; Wayne J.; et al.	June 26, 2014
20140178959	Flow Tube Reactor	Powell; Wayne J.; et al.	June 26, 2014
20140174945	Systems and Methods of Improved Fermentation	Powell; Wayne J.; et al.	June 26, 2014
20140174944	Systems and Methods of Improved Fermentation	Powell; Wayne J.; et al.	June 26, 2014
20140154749	Processing Biomass	Medoff; Marshall	June 5, 2014
20140144078	Modular Automated Aeroponic Growth System	Gonyer; Daegan; et al.	May 29, 2014
20140137472	Controlled environment and method	Anderson; Dana; et al.	May 22, 2014
20140101999	Growing System for Hydroponics and/or Aeroponics	Gardner; Guy M.; et al.	April 17, 2014

(Continued)

TABLE 5.2 (*Continued*)

US Patents Related to Aeroponics Since 2001 by Number, Title, Principal Investigator, and Date of Issue

Patent Number	Title	Principal Investigator	Date of Issue
20140083008	Hydroponic Growing System	Kotsatos; Vasilios M.	March 27, 2014
20140051575	Fungicide Hydroximoyl-Tetrazole Derivatives	Benting; Juergen; et al.	February 20, 2014
20140033609	Expandable Plant Growth System	Tyler; Gregory J.; et al.	February 6, 2014
20140005230	Fungicide Hydroximoyl-Tetrazole Derivatives	Benting; Juergen; et al.	January 2, 2014
20140000162	Aeroponic Growing System and Method	Blank; Timothy A.	January 2, 2014
20130326950	Vertical Agricultural Structure	Nilles; Steven M.	December 12, 2013
20130312136	Methods and Compositions for Modulating Gene Expression in Plants	Liseron-Monfils; Christophe; et al.	November 21, 2013
20130302437	Composition Comprising an Elicitor of the Plant Immune System	Buonatesta; Raffaele; et al.	November 14, 2013
20130289077	Fungicide Hydroximoyl-Tetrazole Derivatives	Benting; Juergen; et al.	October 31, 2013
20130266556	Processing Biomass	Medoff; Marshall	October 10, 2013
20130219979	Plant Nutrient Coated Nanoparticles And Methods for their Preparation and Use	Deb; Nilanjan	August 29, 2013
20130216520	Processing Biomass	Medoff; Marshall	August 22, 2013
20130190353	Fungicide Composition Comprising a Tetrazolyloxime Derivative and a Thiazolylpiperidine Derivative	Coqueron; Pierre-Yves; et al.	July 25, 2013
20130183735	Processing Biomass	Medoff; Marshall	July 18, 2013
20130178447	Pesticide Composition Comprising Fosetyl-Aluminium, Propamocarb-Hcl and an Insecticide Active Substance	Hungenberg; Heike; et al.	July 11, 2013
20130164818	Processing Biomass	Medoff; Marshall	June 27, 2013
20130123305	Fungicide Hydroximoyl-Tetrazole Derivatives	Beier; Christian; et al.	May 16, 2013
20130116287	Fungicide Hydroximoyl-Heterocycles Derivatives	Beier; Christian; et al.	May 9, 2013

(*Continued*)

TABLE 5.2 (*Continued*)

US Patents Related to Aeroponics Since 2001 by Number, Title, Principal Investigator, and Date of Issue

Patent Number	Title	Principal Investigator	Date of Issue
20130081327	Mechanism for Aeration and Hydroponic Growth of Plant Applications	Buck; Jeremiah; et al.	April 4, 2013
20130045995	Fungicide Hydroximoyl-Heterocycles Derivatives	Beier; Christian; et al.	February 21, 2013
20130040993	Fungicide Hydroximoyl-Heterocycles Derivatives	Beier; Christian; et al.	February 14, 2013
20130036669	Apparatus for Growing Plants Hydroponically	Rabii; Mehdi; et al.	February 14, 2013
20130025197	Reduction and Induction of Protein Folding	Tambunga; Gabriel James	January 31, 2013
20130012546	Fungicide Hydroximoyl-Tetrazole Derivatives	Beier; Christian; et al.	January 10, 2013
20120316376	Processing Biomass	Medoff; Marshall	December 13, 2012
20120297678	Vertical Aeroponic Plant Growing System	Luebbers; Terry; et al.	November 29, 2012
20120282345	Pesticide Composition Comprising Propamocarb-Hydrochloride and an Insecticide Active Substance	Van Den Eynde; Koen; et al.	November 8, 2012
20120279126	Apparatus for Aeroponically Growing and Developing Plants	Simmons; Robert S.	November 8, 2012
20120270843	Pesticide Composition Comprising Propamocarb-Fosetylate and an Insecticidally Active Substance	Van Den Eynde; Koen; et al.	October 25, 2012
20120233918	Hydroponics Equipment Cleaning Method	Hayes; Brian; et al.	September 20, 2012
20120230145	Low-Turbulent Aerator and Aeration Method	Ladouceur; Richard	September 13, 2012
20120198763	Photovoltaic Greenhouse Structure	Chuang; Mei-Chen; et al.	August 9, 2012
20120129697	Plant Growth Enhancing Mixture and Method of Applying Same	Stoller; Jerry; et al.	May 24, 2012
20120124905	Method and Apparatus For Growing Plants	Harder; Scott	May 24, 2012
20120094358	Processing Biomass	Medoff; Marshall	April 19, 2012
20120094355	Processing Biomass	Medoff; Marshall	April 19, 2012
20120090236	Aeroponic Plant Growing System	Orr; Gregory S.	April 19, 2012

(Continued)

TABLE 5.2 (*Continued*)

US Patents Related to Aeroponics Since 2001 by Number, Title, Principal Investigator, and Date of Issue

Patent Number	Title	Principal Investigator	Date of Issue
20120090235	Apparatus for a Plant Growth Medium	Horn; Briar; et al.	April 19, 2012
20120085026	Expandible Aeroponic Grow System	Morris; David W.	April 12, 2012
20120077247	Processing Biomass	Medoff; Marshall	March 29, 2012
20120046169	Compositions Comprising a Strigolactone Compound and a Chito-Oligosaccharide Compound for Enhanced Plant Growth and Yield	Dahman; Peter; et al.	February 23, 2012
20120031505	Irrigation System and Method	Jensen; Jarl	February 9, 2012
20120027741	Pesticide Composition Comprising a Tetrazolyloxime Derivative and a Fungicide or an Insecticide Active Substance	Coqueron; Pierre-Yves; et al.	February 2, 2012
20110296757	Portable Hydroponic Terrace Cart	McGrath; Kevin Robert	December 8, 2011
20110294829	Fungicide Hydroximoyl-Heterocycles Derivatives	Beier; Christian; et al.	December 1, 2011
20110294752	Triterpene Compositions and Methods for Use Thereof	Arntzen; Charles J.; et al.	December 1, 2011
20110287108	Pesticide Composition Comprising a Tetrazolyloxime Derivative and a Fungicide or an Insecticide Active Substance	Coqueron; Pierre-Yves; et al.	November 24, 2011
20110201613	Fungicide Hydroximoyl-Tetrazole Derivatives	Beier; Christian; et al.	August 18, 2011
20110190353	Fungicide Hydroximoyl-Tetrazole Derivatives	Beier; Christian; et al.	August 4, 2011
20110146146	Method and Apparatus for Aeroponic Farming	Harwood; Edward D.	June 23, 2011
20110120005	Hydroponic Plant Growth Systems with Activated Carbon and/or Carbonized Fiber Substrates	King; Mark J.; et al.	May 26, 2011
20110105566	Fungicide Hydroximoyl-Tetrazole Derivatives	Beier; Christian; et al.	May 5, 2011
20110081336	Processing Biomass	Medoff; Marshall	April 7, 2011
20110081335	Processing Biomass	Medoff; Marshall	April 7, 2011

(*Continued*)

TABLE 5.2 (*Continued*)
US Patents Related to Aeroponics Since 2001 by Number, Title, Principal Investigator, and Date of Issue

Patent Number	Title	Principal Investigator	Date of Issue
20110061297	Apparatus for Aeroponically Growing and Developing Plants	Simmons; Robert S.	March 17, 2011
20110061296	Apparatus for Aeroponically Growing and Developing Plants	Simmons; Robert Scott	March 17, 2011
20110056132	Growing System for Hydroponics and/or Aeroponics	Gardner; Guy	March 10, 2011
20110052555	Pesticide Composition Comprising a Tetrazolyloxime Derivative and a Fungicide or an Insecticide Active Substance	Coqueron; Pierre-Yves; et al.	March 3, 2011
20110034445	Fungicide Hydroximoyl-Heterocycles Derivatives	Beier; Christian; et al.	February 10, 2011
20110023359	Aeroponic Growing Apparatus and Method	Raring; David	February 3, 2011
20110021572	Fungicide Hydroximoyl-Tetrazole Derivatives	Beier; Christian; et al.	January 27, 2011
20110016782	Method And Apparatus for Growing Plants	Harder; Scott	January 27, 2011
20110015236	Fungicide Hydroximoyl-Tetrazole Derivatives	Beier; Christian; et al.	January 20, 2011
20100286173	Fungicide Hydroximoyl-Tetrazole Derivatives	Beier; Christian; et al.	November 11, 2010
20100269409	Tray for Hydroponics Growing of Plants and Hydroponics Tank Having the Tray	Johnson; Paul	October 28, 2010
20100236164	Photovoltaic Greenhouse Structure	Chuang; Mei-Chen; et al.	September 23, 2010
20100218423	Aeroponic Plant Growing System	Walhovd; Zack Allen	September 2, 2010
20100189824	Triterpene Compositions and Methods for Use Thereof	ARNTZEN; CHARLES J.; et al.	July 29, 2010
20100152270	Pesticide Composition Comprising a Strigolactone Derivative and a Fungicide Compound	Suty-Heinze; Anne; et al.	June 17, 2010
20100144534	Compositions and Methods for the Control of Nematodes and Soil Borne Diseases	Pullen; Erroll M.	June 10, 2010
20100137373	Pesticidal Composition Comprising a Strigolactone Derivative and an Insecticide Compound	Hungenberg; Heike; et al.	June 3, 2010

(Continued)

TABLE 5.2 (*Continued*)

US Patents Related to Aeroponics Since 2001 by Number, Title, Principal Investigator, and Date of Issue

Patent Number	Title	Principal Investigator	Date of Issue
20100130357	Pesticidal Composition Comprising a Pyridylethylbenzamide Derivative and an Insecticide Compound	Hungenberg; Heike; et al.	May 27, 2010
20100124583	Processing Biomass	Medoff; Marshall	May 20, 2010
20100113278	Pesticidal Composition Comprising Synthetic Compound Useful as Nodulation Agent of Leguminous Plants and a Fungicide Compound	Suty-Heinze; Anne; et al.	May 6, 2010
20100101145	Hydroponic System	Bergen; Klaes Alexander	April 29, 2010
20100063143	Pesticide Composition Comprising Propamocarb-Hydrochloride and an Insecticide Active Substance	van Den Eynde; Koen; et al.	March 11, 2010
20100063007	Pesticide Composition Comprising Fosetyl-Aluminum and an Insecticide Active Substance	Hungenberg; Heike; et al.	March 11, 2010
20100048404	Pesticidal Composition Comprising a Synthetic Compound Useful as Nodulation Agent of Leguminous Plants and an Insecticide Compound	Hungenberg; Heike; et al.	February 25, 2010
20100035844	Pesticide Composition Comprising Propamocarb-Fosetylate and an Insecticidally Active Substance	Van Den Eynde; Koen; et al.	February 11, 2010
20100035840	Pesticidal Composition Comprising Fosetyl-Aluminium, Propamocarb-Hcl and an Insecticide Substance	Hungenberg; Heike; et al.	February 11, 2010
20100029776	Pesticidal Composition Comprising Fenamidone and an Insecticide Compound	Hungenberg; Heike; et al.	February 4, 2010
20090305888	Materials and Methods for Providing Oxygen to Improve Seed Germination and Plant Growth	Li; Yuncong; et al.	December 10, 2009
20090293357	Aeroponic Atomizer for Horticulture	Vickers; Ross; et al.	December 3, 2009
20090281151	Pesticidal Composition Comprising a 2-Pyrdilmethylbenzamide Derivative and an Insecticide Compound	Hungenberg; Heike; et al.	November 12, 2009
20090151248	Devices and Methods for Growing Plants	Bissonnette; W. Michael; et al.	June 18, 2009
20090120091	Power Generation System	DuBois; John R.	May 14, 2009

(*Continued*)

TABLE 5.2 (*Continued*)
US Patents Related to Aeroponics Since 2001 by Number, Title, Principal Investigator, and Date of Issue

Patent Number	Title	Principal Investigator	Date of Issue
20080295400	Method and Apparatus for Aeroponic Farming	Harwood; Edward D.; et al.	December 4, 2008
20080282610	Devices and Methods for Growing Plants	Bissonnette; W. Michael; et al.	November 20, 2008
20080245282	Dispersion and Aeration Apparatus for Compressed Air Foam Systems	Richards; William Henry	October 9, 2008
20080222949	Devices and Methods for Growing Plants	Bissonnette; W. Michael; et al.	September 18, 2008
20080155894	Soil-less Seed Support Medium and Method for Germinating a Seed	Bissonnette; W. Michael; et al.	July 3, 2008
20070271842	Systems and Methods for Controlling Liquid Delivery and Distribution to Plants	Bissonnette; W. Michael; et al.	November 29, 2007
20070271841	Devices and Methods for Growing Plants	Bissonnette; W. Michael; et al.	November 29, 2007
20070260039	Methods of Producing Silk Polypeptides and Products Thereof	Karatzas; Costas N.; et al.	November 8, 2007
20070102359	Treating Produced Waters	Lombardi; John A.; et al.	May 10, 2007
20060179711	Devices and Methods for Growing Plants	Bissonnette; W Michael; et al.	August 17, 2006
20060168881	Hydroponic Plant Nutrient Kit and Method of Use	Straumietis; Michael James	August 3, 2006
20060148732	Inhibition of NF-kappaB by Triterpene Compositions	Gutterman; Jordan U.; et al.	July 6, 2006
20060096168	Method and a Kit for Shaping a Portion of a Woody Plant into a Desired Form	Golan; Ezekiel	May 11, 2006
20060073222	Triterpene Compositions and Methods for Use Thereof	Arntzen; Charles J.; et al.	April 6, 2006
20060053691	Method and Apparatus for Aeroponic Farming	Harwood; Edward D.; et al.	March 16, 2006
20050257424	Devices and Methods for Growing Plants	Bissonnette, W. Michael; et al.	November 24, 2005
20050246955	Devices and Methods for Growing Plants	Bissonnette, W. Michael; et al.	November 10, 2005
20050246954	Devices and Methods for Growing Plants	Bissonnette, W. Michael; et al.	November 10, 2005
20050241231	Methods and Devices for Promoting the Growth of Plant Air Roots	Bissonnette, W. Michael; et al.	November 3, 2005

(*Continued*)

TABLE 5.2 (*Continued*)

US Patents Related to Aeroponics Since 2001 by Number, Title, Principal Investigator, and Date of Issue

Patent Number	Title	Principal Investigator	Date of Issue
20050198897	Liquid Fractionation System Useful for Growing Plants	Wainright, Robert E.; et al.	September 15, 2005
20050108938	Method for the Propagation of and Aeroponic Growing of Plants and Vessels Therefore	Bakula, Roxanne E.; et al.	May 26, 2005
20050102895	Soil-less Seed Support Medium and Method for Germinating a Seed	Bissonnette, W. Michael; et al.	May 19, 2005
20050054830	Methods and Apparatus for Spinning Spider Silk Protein	Islam, Shafiul; et al.	March 10, 2005
20050011118	Seed Germination and Plant Supporting Utility	Umbaugh, Raymond E. JR.	January 20, 2005
20040102614	Methods and Apparatus for Spinning Spider Silk Protein	Islam, Shafiul; et al.	May 27, 2004
20040055213	Low Pressure Aeroponic Growing Apparatus	Wainwright, Robert E.; et al.	March 25, 2004
20030203049	Triterpene Compositions and Methods for Use Thereof	Arntzen, Charles J.; et al.	October 30, 2003
20030088888	Method of Plant Transformation	Vainstein, Alexander; et al.	May 8, 2003
20030054052	Triterpene Compositions and Methods for Use Thereof	Haridas, Valsala; et al.	March 20, 2003
20030039705	Triterpene Compositions and Methods for Use Thereof	Arntzen, Charles J.; et al.	February 27, 2003
20030031738	Triterpene Compositions and Methods for Use Thereof	Haridas, Valsala; et al.	February 13, 2003
20020132021	Elicited Plant Products	Raskin, Ilya; et al.	September 19, 2002
20010052199	Stackable Planter	Klein, Shoshana; et al.	December 20, 2001

REFERENCES

Aerogrow website, accessed 4/17/20, www.aerogrow.com.
CropScience website, accessed 4/17/20, www.cropscience.bayer.com.
Living Greens website, accessed 9/8/19, www.livinggreensfarm.com.
Lynch, C., 2018, *Farm Flavor, How Walt Disneyworld's Farm Grows the Most Magical Produce on Earth*, 5//2/18, www.farmflavor.com.
Spinoff, 2007*Progressive Plant Growing Is a Blooming Business*, 4/23/7, p. 64, www.nasa.gov/vision/earth/technologies/aeroponic_plants.html.
Xyleco website, accessed 4/17/20, www.xylexo.com.

6 Aeroponic Business

Achieve the greatest volume and highest quality of produce possible, while reducing operating costs, and maximizing your profitability by growing smart.

Tom Blount, Expert at US Hydroponic Association

The business of growing food using aeroponics is limited to a small number of start-up companies. These ventures employ different variations on the aeroponic system and many have patents on their technology (see Chapter 5). Some use vertical columns and trickle down nutrient solution similar to hydroponics nutrient flow technique (NFT) only going vertical. Others are using the traditional misting approach either in conjunction with vertical columns or germinating seeds in a fabric matrix or using horizontal trays stacked vertically. The term vertical farming needs to be clearly defined since it can have multiple meanings. Vertical columns or walls are one method that can be described as vertical farming. However, stacking horizontal trays with layers of LED lights for each layer in a vertical configuration is the alternative version of vertical farming. So, this chapter includes initially a description of the most successful pioneering work in this area, the Tower Garden, and its proliferation around the US. It then describes a number of other companies, both manufacturers and growers, who are selling or using aeroponic systems, some in greenhouses and others indoors, to grow food. These companies are located in the US and around the world. The companies are described in alphabetical order for reference purposes after several examples of the Tower Garden systems are given.

The success of aeroponic business activity in the last 20 years can be attributed in a large extent to the innovative work done at Disneyworld in the 90's and specifically the entrepreneurial efforts of Tim Blank and Tower Gardens at the beginning of the 21st century (Tower Garden, 2019). These simple units that can be used by hobbyists, by academic institutions, by restaurants, and many other venues have opened the door for this technology to get exposure and mature. Blank has expanded this technology by selling hundreds of these systems. Here are some examples of what these systems are being used for.

TOWER GARDENS (WWW.TOWERGARDENS.COM)

Tower Garden makes a vertical aeroponic growing system with a nutrient reservoir with 20 gallons and towers that can grow 20–40 plants. Figure 6.1 is a picture of two Tower Gardens with plants growing on them. The reservoir contains a submerged pump that pumps the nutrient solution to the top of the tower and then it flows inside the tube and back to the reservoir, irrigating the roots of the plants growing on the outside of the tower.

FIGURE 6.1   Tower Garden vertical systems.

ALASKA

Andrew Hohenthaner is an indoor gardener who enjoys the science of aeroponics (Aeroponic blog, 2015). Hohenthaner gardens in a small (12-by-15-foot) greenhouse in his back yard in Juneau. In 2009, he decided to purchase his "top-of-the-line" aeroponic system, which is expandable and has room to grow 36 plants. Hohenthaner's system looks like a science project, with tubes and pumps coming out of plastic trays. There are pumpkin flowers growing from succulent vines. The pumpkins didn't germinate the first year, but he did grow cilantro, oregano, basil, habanera peppers, lettuce, tomatoes, sugar peas, and cucumbers. He plans to grow strawberries and try the pumpkins again. During the winter, he plans to move the entire system into his garage and keep growing. There is no weeding to be done, no pesticides needed, and no working in the rain. His first plants grew in 38 days and he is now growing a second crop of plants. Hohenthaner said that his basil plants are producing three pounds of basil per week.

ARIZONA

In 2014, Kathleen Rose began selling an aeroponics system with Juice Plus (Acoba, 2014). "I've always wanted a garden," says Rose, a Foothills resident who sells nutritional supplements and the Tower Garden for the company. "But I never had time or space." The 4.5-foot Tower Garden with seeds and plant food costs $525. Because water circulates, aeroponic gardening uses far less water than in-ground plants, Rose says. She's growing kale, celery, zucchini, cucumber, cherry tomatoes, herbs and nasturtium in the standard tower and a 6-foot model. Any type of plant that doesn't grow on a tree or in the soil, such as root vegetables, can be grown in an aeroponics system.

CALIFORNIA

In 2017, UCLA installed 50 aeroponic towers on campus (Bazak, 2017). Each tower has up to 44 growing sites and will help supply leafy greens to Bruin Plate, the university restaurant. The plants produce new greens every 4 weeks. The towers use about 90% less land and water than conventional farming. They each have a reservoir for plant roots to hang inside and a pump that recycles the water. The process better oxygenates the plant roots and speeds up plant growth. Similar towers are already located outside the Stewart and Lynda Resnick Neuropsychiatric Hospital. Christina Lee, the undergraduate student government student wellness commissioner and member of UCLA's Healthy Campus Initiative, said she thinks the project will be educational for students. "It is definitely a useful project," Lee said. "It might be inspiring to students to see that sustainability can be achieved in an urban landscape, and that something like this can also produce more food. So it is also useful for food security in a broader aspect."

COLORADO

Montrose resident, Doreen Dwyer, has a Tower Garden in her home. She purchased the tower and planted it in 2015 (Wright, 2015). She already has been growing lettuce, peppers, parsley, tomatoes, and even strawberries. It may sound like science fiction to garden without dirt, no weeds, with about 1/10th of the water normally utilized and never sit on your knees to tend to plants. However, it is successfully being done through aeroponics and vertical gardening. "We eat a green salad every evening," Dwyer said. "And the taste of the produce that comes off this (tower) is so much better." Dwyer also said everything she has grown tastes richer, "more like it should."

Riverhouse Children's Center, with support from Alpine Bank, in 2017 launched a program to grow year-round organic fruits and vegetables through a soil-free method called aeroponics in Durango (Andersen, 2017). The program involves four "tower gardens" to be grown at the school and provide Science, Technology, Engineering and Mathematics programming to students.

FLORIDA

In a city clustered with high-rise condos, Arthur Chernov and his wife Cheyenne live in one of the few pockets of houses in Aventura. Although fortunate to have a backyard, the Chernovs prefer aeroponics over traditional gardening (Devaney, 2015). This method eliminates soil and uses only water and nutrients to grow edible fruits and vegetables. The Chernovs like to use the Tower Garden they purchased in 2015, a system that can grow up to 20 different plants vertically, reducing the amount of space that a regular backyard garden would use. The system requires electricity for the built-in fountain that distributes the nutrient and water mix to the plants, as well as a minimum of 4 h of sunlight daily. Some of the pros of aeroponic gardening are that the plants are pesticide-free, grown locally, and the technology reduces the chance of pesticide and nutrient runoff from contaminating the land, air, and sea.

Arthur is pushing to get Tower Gardens into the local schools, so that students of all academic levels can learn the science behind aeroponics and the importance of growing their own food. For the 2014–2015 school year, the Sunny Isles Beach K-8 School Fund purchased six Tower Gardens, giving three each to both the Norman S. Edelcup Sunny Isles Beach K-8 School and Alonzo & Tracy Mourning Senior High School. "The kids loved getting involved with their hands with growing and doing," Arthur said. "If you get kids young enough that have an open mind and are willing to listen, you can change a generation."

Illinois

O'Hare Airport uses 11' Tower Gardens to grow herbs and veggies. In 2011, Sarah Gardner stated, "Yep, it's right between Terminals 2 and 3 on Concourse G: A no-soil, vertical garden that grows everything from Swiss Chard to green beans" (CDA, 2011).

Chicago's aviation commissioner Rosemarie Andolino became enamored with these "aeroponic" gardens after seeing one on an episode of "Nightline." She asked airport concessionaire HMS Host to fund one at O'Hare. Now some of O'Hare's restaurants are serving the garden's veggies to customers. Brad Maher, HMS host stated, "It's very functional. I was very surprised at the amount of crops that we actually harvest out of here. We get a full crop every 4 weeks. O'Hare's new garden looks pretty too. Maybe, O'Hare hopes, pretty enough to take your mind off that flight delay."

Dr. John Saran and his wife, Janet, of Naperville purchased an aeroponic garden for their business and home (Reality Fitness, 2014). I am hoping that our grandchildren will have memories of our "garden" and the fresh produce we picked together for our meals. It will be different from my grandparents' gardens, but I am excited to share with them the miracle of watching our seeds grow into food for our table. Many aeroponic garden users choose to bring their gardens in the house for the winter months, and by adding lights, they can continue to have fresh vegetables and grow flowers all year long.

Indiana

Liberty Christian School Garden Club in Anderson started in 2008. "Our science teachers are real excited to use this in the classroom," she said (Bibbs, 2018). Though science instruction doesn't really start till first grade at Liberty Christian, kindergarten teacher Tammy Ramsey said she's excited about the aeroponics tower. One Tower Garden was purchased in 2018. Logan Parker, 11, helped stack the pots labeled A, B and C onto the 20-gallon water tank of the aeroponics garden tower in the hallway at Liberty Christian School. "The base was already assembled, so all we had to do was put in the rods and the pots," the fifth-grader said. "Before we did this, we watched two videos on it, how it would look in the end and how we put it together. It was fun, and I don't think I'll ever forget it." The aeroponics tower, where students will grow flowers, fruits and vegetables, is the first activity of several the school's garden club plans this year. Cory Bohlander, president of Madison County Farm Bureau, said his organization was excited to donate the funds to purchase the tower, which has a base cost of $500, not including the grow lights and additional pots.

LOUISANA

Rouses Markets installed in 2012 a sustainable aeroponic rooftop garden above a downtown New Orleans store—New Roots on the rooftop urban farm was the first in the country (Rouses Market, 2012). Parsley, basil and cilantro are among the herbs the company is growing to package and sell on the building's ground floor. Rouses Markets is the first grocer in the country to develop its own aeroponic urban farm on its own rooftop, says managing partner Donny Rouse. And they could not have picked a more picturesque location. "The flat rooftop on this store is perfect for urban farming," says Rouse. "And the view of downtown is postcard-perfect. I imagine we will do a lot of dinners up here on the farm." Rouses Markets downtown store sits just blocks from the Superdome, French Quarter, and Mississippi River. The vertical aeroponic Tower Garden uses water rather than soil, and allows you to grow up instead of out. Chef Louis "Jack" Treuting, Rouses Culinary Director, first saw Roots on the Rooftop as a way to provide fresh herbs for the food Rouses chefs prepare, but quickly saw potential to expand the program to include retail. "I knew if our chefs wanted it, so would our customers."

NEW YORK

SUNY Potsdam brought Tower Garden farms on campus in 2017 that grow greens in mid-air: Plastic pillars allow plants to thrive without soil and water; uses a system known as aeroponics (Kenmore, 2017). At SUNY Potsdam's new greenhouse, college students, elementary school kids and local residents are learning how to grow greens—in mid-air.

Bronx Green Machine was established in New York City public schools to grow food. Probably the most unique factor is that the towers do not use soil or water as a growing medium. Instead, the plants are suspended in air, with water and nutrients dripped onto the roots, a system known as aeroponics (Green Bronx Machine, 2017). "Everybody thinks of growing things—things need soil," said Mr. McCarthy. "But not everything needs soil. The students, all in sixth or seventh grade, assembled and planted two of the nine towers completed during the workshop." "Ray (Bowdish) and I have been working on bringing stuff the students grow on campus to the dining services," said Ms. Conger. She plans on using four of the Tower Farms to grow greens and herbs in buildings around campus, so students can see their food being grown. "We're hoping it will get to the point where we have outdoor raised beds the dining service can use," said Ms. Conger. "But this is a great first step."

OKLAHOMA

Scissortail Farms in Tulsa run by John Sulton and his partner, Rob Walenta, operate Scissortail Farms on a 7.5-acre site west of Tulsa (Smith, 2014). In recent weeks they started growing 50 varieties of lettuce, including kale, chard, mustard greens and arugula in their 26,800-square-foot, state-of-the-art greenhouse. "It is nice to have everything indoors," Sulton said. Juniper Restaurant in Tulsa buys from Sulton. "We buy from Scissortail Farms because their plants are grown locally; the quality is there," said Todd Phillips, general manager of Juniper. Sulton and Walenta employ aeroponic farming to grow the vegetables. In aeroponics, farmers grow plants without soil and with little water. They use pumps and gravity to feed the roots with water loaded with nutrients.

The method is more expensive than traditional farming, but customers prefer the produce, Sulton said. Phillips said Scissortail's prices are higher than buying vegetables from a traditional farm. "A lot higher," Phillips said. "However, the product has a much longer shelf life. It has the freshness, flavor, and the vitamins."

The growing method allows Scissortail Farms to consistently provide fresh produce within 24 hours of harvest, rather than waiting several days for shipment from outside the community, Sulton said. The process also eliminates the need for fertilizer or manure. The thousands of plants growing out of 11-foot-tall white aeroponic tubes are watered every 5 min with conditioned water, Sulton said. Each tower has 44 slots for growing plants. A few towers are 13-feet tall. "Altogether, we've got a little over 62,000 individual slots for plants at any given time."

START-UP AEROPONIC COMPANIES

In the past 10 years there have been many companies formed providing aeroponic growing systems. The key businesses described in this chapter include 12 manufacturers (Aero Development, AssenceGrows, Agrihouse, Cloudponics, Green Hygenics Holdings, Gro-pod, Helioponics, Podplants, and Treevo); 16 Growers (Aerofarms, Aero Spring Gardens, Agricool, 80 Acre Farms, Farmedhere, Grow Anywhere, Growx, Indoor Farms of America, Indoor Harvest Corp, James E. Wagner Cultivation, Just Green, Lettuce Abound Farms, Living Greens, Plenty, Riviera Creek, and True Garden), and one Manufacturer and Grower (Amplified Ag Inc.) These companies are involved with a variety of different systems, applications, and designs. Each of these companies is listed in Table 6.1 and described below alphabetically. Two companies have gone out of business in the past couple of years—FarmedHere and Indoor Farms of America.

AERO DEVELOPMENT CORP (WWW.THINKAERO.CO)

Manufacturer of Vertical Commercial and R&D Column Systems—Pennsylvania (Aero Development website, www.thinkaero.co)

Since 2011, Aero Development Corp has been an early pioneer in the design, development, and testing of Aeroponic growing systems, capable of being scaled for large, commercial installations. Their unique, patented, vertical growing systems

TABLE 6.1

Aeroponic Companies

Company	Type of Business	Location	Type of Aeroponics
Aero Development Corp	Manufacturer	PA, USA	Vertical
AeroFarms	Grower	NJ, USA	Stacked misted
Aero Springs Garden	Grower	Singapore	Stacked
Aessense Grows	Manufacturer	CA, USA	Vertical stacked
Agricool	Grower	France	Vertical walls
Agrihouse	Manufacturer	CO, USA	Vertical walls
Amplified Ag	Mfgr/Grower	SC, USA	Stacked/Controllers
Cloudponics	Manufacturer		Controllers
80 Acres	Grower	OH, USA	Vertical
Green Hygenics Holdings	Manufacturer		Hybrids
Grow Anywhere	Grower	CO, USA	Vertical misted
Gro-Pod	Manufacturer	UK	Education
GrowX	Grower	Netherlands	Horizontal
FarmedHere	Grower		
Helioponics	Manufacturer	IN, USA	Vertical
Indoor Farms of America	Grower	NV, USA	Vertical
Indoor Harvest Corp	Grower	TX, USA	Custom
Just Green	Grower	NY, USA	Stacked misted
JWC Inc	Grower	Canada	Horizontal
Lettuce Abound	Grower	MN, USA	Vertical
Living Greens	Grower	MN, USA	Vertical
Plenty	Grower	CA, USA	Vertical
PodPlants	Manufacturer	Australia	Vertical
Riviera Creek	Grower	OH, USA	Horizontal Misted
Treevo	Manufacturer	Israel	Tree designs
True Garden	Grower	AZ, USA	Vertical

have a proven track record of producing a virtually unlimited variety of high caliber, chemical, and pesticide-free produce. Growing vertically allows the generation of high volumes of produce in less space compared to conventional soil-based growing as well as hydroponic horizontal growing systems. Their unique closed loop irrigation system requires less than 10% water than soil-based growing, and without the algae challenges faced by other water-based growing systems. Their cloud-based monitoring system provides comprehensive, real-time data on every aspect of the growing environment, as well as tracking all the back office operating components from seed to sale.

They manufacture five different aeroponic growing systems from ones that can be used in the kitchen to commercial units. These include:

- AERO Kitchen Garden which comes with a bonus starter kit, which includes everything needed to start growing vegetables, herbs, and fruits at home. This system can grow 18 plants in less time than it takes in soil. It measures 8″ wide × 32″ high. This unit is capable of growing fruits and almost any vegetable, herb, or flower (Figure 6.2).

- AERO Mini Mobile Garden which is a versatile growing system designed to bring a fresh harvest of vegetables, herbs, and fruits to the kitchen from the convenience of a patio or porch.
- AERO Midi Mobile Garden (indoor/outdoor) which is on caster wheels has grow cups for up to 36 plants. It measures 26″ wide × 57″ high × 13″ deep (Figure 6.3).
- AERO Midi Mobile Garden which is similar to the Mini except it has grow cups for up to 54 plants. It measures 50″ wide × 57″ high × 13″ deep (Figure 6.4).
- AERO Mobile Garden which has three towers and has grow cups for 72 plants.

FIGURE 6.2   AERO kitchen garden.

FIGURE 6.3   AERO mini mobile garden.

FIGURE 6.4 AERO midi mobile garden.

The fifth system is the commercial inline system that can be installed in a greenhouse or a warehouse and consists of 8 ten-foot tall columns arranged on a 6-foot diameter base pod with a capacity of 608 plants in this footprint (approximately 20 plants per square foot). In a greenhouse at Garden Spot Retirement Village in New Holland, PA, is a 4,600 square foot greenhouse that has 25 pods with a capacity of 15,000 plants. The inline system is the newest generation of aeroponic vertical column and it increases the capacity by almost three times the pod systems can achieve. In a 4,600 square foot greenhouse, these systems could grow approximately 43,000 plants.

AEROFARMS (WWW.AEROFARMS.COM)

Grower Stacked Horizontal Trays Indoors—New Jersey (Aerofarms website)

AeroFarms based in Newark, NJ recently started its ninth farm, billed as the world's largest vertical farm, at its new global headquarters, and it has others in development in multiple US states and in four continents. The initial farm was built in 2014 in downtown Newark, NJ inside a former nightclub and lounge. The farm consists of a 15-foot-high stack of planters, each one 10 feet long, 3 feet wide, and a foot or 2 deep. There is not any soil beneath the leafy green vegetation bursting from the flower boxes. There is only air (Figure 6.5).

Essentially, an aeroponic farmer sprays a mist of a high-nutrient solution on plants to make them grow. The process takes far less space and water than nature would require, and zero pesticides. Until recently, technological limits kept true aeroponics largely beyond the reach of commercial growers.

AeroFarms founder Ed Harwood says cheaper, more advanced equipment, from software to lights to fabric, has made aeroponics feasible. AeroFarms raised more

FIGURE 6.5   Aerofarms indoor vertical farm.

than $36 million in venture capital. According to an article in Bloomberg Business 2014, AeroFarms planned to expand to 70 employees from 15 and build an urban farm capable of turning out 1.5 million pounds of produce per year, enough to supply about 60,000 people (Friedman, 2014).

For a crop such as kale or arugula, Harwood says his setup can grow plants faster with as little as 10% of the water. AeroFarms begins by scattering seeds on a permeable microfleece cloth stretched over a planter. Inside the modular, stackable container, nozzled hoses pump oxygen and the nutrient mist onto the fabric, forming a membrane on which the seeds germinate. Eventually they sprout silky roots that poke down through the micro-fleece to absorb more of the mist. As the plants grow, additional hoses feed them carbon dioxide as sensors monitor humidity and room temperature. A conveyor belt propels the greens through a trimming device and discards the roots from the microfleece so the fabric can be washed and reused.

Harwood, who holds a doctorate in dairy science from the University of Wisconsin at Madison, was the associate director of the Cooperative Extension at Cornell, which links up with local farmers to research agriculture and food systems. There, in 2002, he built a prototype aeroponic system with microfleece after testing other textiles. "I became friends with all the ladies at JoAnn Fabrics," he says. The following year he launched farming venture GreatVeggies, but he couldn't get it off the ground. In 2009, he received $500,000 in funding from 21 Ventures and the Quercus Trust to start AeroFarms.

The company's plan was to sell aeroponic equipment, but as it conducted more trials—and lights and other equipment grew radically cheaper—growing vegetables began to seem more profitable, says Chief Executive Officer David Rosenberg, adding that produce prices will be similar to those of typical large field growers. AeroFarms has focused on the multibillion-dollar market for leafy-green salad ingredients, including fresh herbs and specialty microgreens.

In 2015, The New York Times published an article entitled Prime Newark Farmland, Inside an Old Steel Factory (Hughes, 2015). A former Grammer, Dempsey, and Hudson steel plant in the Ironbound section of Newark is being razed by the RBH Group to make way for a giant custom-built complex for its sole tenant, AeroFarms, a company producing herbs and vegetables in an indoor, vertical environment. Instrumental in reviving parts of Newark, the RBH Group sees the venture as a way to create jobs, clear a shabby block, and supply a healthy, locally grown food source. The complex, a group of metal-block, low-slung buildings, some connected, some not, also has prominent backers. Through its Urban Investment Group, Goldman Sachs is picking up the bulk of the $39 million cost for development of the AeroFarms Ironbound complex, using equity, debt and bridge financing. Prudential Financial, whose headquarters are now in Newark, is also an investor. The project has been awarded $9 million in city and state money, in tax credits and grants. This 69,000-square-foot complex will also contain labs, offices, and a cafe and was finished in 2016.

Unlike urban vegetable gardens of the past that took advantage of empty lots or evolved in rooftop greenhouses, AeroFarms employs so-called aeroponics and stacks its produce vertically, meaning plants are arrayed not in long rows but upward. Because the farming is completely indoors, it relies on LED bulbs, with crops growing in cloth and fed with a nutrient mist.

Critics of vertical farming have complained that taste can suffer when food is cultivated without soil or sun, while proponents say vertical farms are extremely efficient and have a small environmental impact. They take up minimal space, grow round the clock, and are near the markets that sell their crops, reducing the need for long truck trips. Vertical farms are also far less susceptible to the vagaries of unpredictable weather like droughts or floods.

"We can deliver anything the plant wants, when it wants it, how it wants it and where it wants it," said David Rosenberg, chief executive of AeroFarms. AeroFarms projects it will reap up to 30 harvests a year, or 2 million pounds of greens, including kale, arugula, and romaine lettuce, Mr. Rosenberg said. At that output, AeroFarms would be among the most productive vertical farms in the country, analysts say. But in an industry where profitability is elusive, success is hardly guaranteed. Indeed, AeroFarms is still lining up customers, which ideally will include grocery chains, schools, and restaurants.

Comparing vertical farms can be tricky. Unlike AeroFarms, some sell whole plants, or by-products like juice and salad dressing. Also, because the height of rooms in vertical farms is often more important than their width, floor measurements can be misleading, some farmers say. Still, in real estate terms, the Ironbound operation would be among the country's largest. About 2/3 of the complex, or 46,000 square feet, will be dedicated to crops, according to the company, in rooms with lofty 30-foot ceilings.

In contrast, FarmedHere, an Illinois company, has grown plants in about 47,000 square feet of a low-slung 93,000-square-foot former box factory near Chicago Midway International Airport. Founded in 2011, FarmedHere was selling its produce to nearly 50 Whole Foods markets, plus other grocery stores, said Mark Thomann, the chief executive (Hughes, 2015).

But in a sign of the risks inherent to the industry, other fledgling companies trying to grow crops in small spaces have sputtered and failed. For example, Alterrus Systems, maker of the shelf-like VertiCrop system, with a greenhouse-like farm on a sun-dappled roof of a parking garage in Vancouver, British Columbia, declared bankruptcy in 2015.

AeroFarms sued FarmedHere in 2014, alleging the Midwestern company stole trade secrets and infringed patents by refusing to pay licensing fees after leasing Harwood's proprietary equipment (Hughes, 2015).

Perhaps the most important element is an array of light-emitting diodes (LEDs). A control panel built by the Lighting Research Center at Rensselaer Polytechnic Institute lets Harwood fine-tune the color wavelength and intensity of the light, which, he says, alter crop yields, taste, appearance, and nutrient levels. Small changes can make kale taste sweeter or arugula more peppery. Lighting is among the biggest costs for an aeroponic grower. Neil Mattson, a professor at Cornell's School of Integrative Plant Science who worked closely with Harwood to develop his "light recipes," says using only red and blue LEDs cut energy costs by 15% (AgFunder, 2014). AeroFarms has benefited from an 85% fall in the price of LEDs, even as the output per watt doubled. That's a huge advantage over earlier efforts to commercialize aeroponics.

In 2016, Diana Swain reported on CBS TV about Aerofarms in a segment entitled "No sunlight, no soil, no problem: Vertical farms take growing indoors." CEO of Aerofarms, David Rosenberg stated that this is fully controlled aeroponic agriculture. It allows us to understand plant biology in ways that humans have never achieved. This vertical farm used LED lights instead of the sun, and cloth instead of soil. The staff at Aerofarms includes electrical engineers, structural engineers, mechanical engineers, as well as plant scientists, crop physiologists, and plant pathologists.

There are farms across Canada trying similar concepts, trying to extend growing seasons and bring fresh produce to cold climates. Experts say the sales pitch for vertical farms is compelling, capitalize on the local food trends like growing plants in dense urban areas. AeroFarms says their system uses less water and no pesticides, produces more plants, more often - all without the waste that comes with industrial agriculture.

But experts say there is a big obstacle. Bruce Bugbee at Utah State University is concerned about the cost of electrical energy. His chief objection is the exorbitant power requirement for such a vertical structure. Plants on the lower floors would require artificial light year-round or expensive mechanical systems to get more light to them. And during a typical winter in northern US cities, he said, average sunlight is only 5%–10% of peak summer levels due to sapped intensity and shorter days (Nelson, 2007). He believes that sunlight is still the gold standard for nutritional quality. AeroFarms says it has cut energy use by focusing on specific types of light. They have shown that the red and blue light is actually how plants see light. And so they believe they have more effective photosynthesis growing indoors than one would have with the natural sun. The company says they're also growing the most cost-effective plants, only microgreens, no tomatoes or cucumbers here.

The $30-million project has attracted big-name investors like Goldman Sachs, and even interest from the Canadian Produce Marketing Association. AeroFarms is currently building the world's biggest vertical farm nearby, with stacks set to be twice as high as what they have now. It's also winning over skeptics. Marion Nestle is a professor of nutrition at NYU and has written numerous books on food and public health. She was recently invited on a tour of the AeroFarms' facility. "Why are we doing this when you can grow vegetables really beautifully in soil? What's the problem here?" She became a convert after tasting the product that goes out to local grocery stores, schools, and farmers' markets. Marion Nestle, Professor of Nutrition, Food Studies and Public Health said, "They taste as good. They really do. In fact, they taste a lot better than the microgreens we get here in New York that sometimes have been on the road for a week."

In 2018, a Dell computer publication had an article entitled "Harnessing IoT to Combat Food Insecurity, Waste and Spoilage" (Aerofarms blog, 2018). Dell is working with 2 of its customers, AeroFarms and IMS Evolve, addressing these problems by using innovative IoT (Internet of Things) technologies to transform food production and delivery. AeroFarms collects millions of data points from farms stacked vertically. Using modern imaging, big data, and machine learning, they're turning everything about a plant into data. According to the article, AeroFarms has developed an indoor vertical farming system that grows food with far fewer resources and less waste than conventional methods. In fact, AeroFarms has achieved 390 times greater productivity than field farming while using 95% less water.

At 70,000 square feet, AeroFarms' flagship facility in Newark, N.J., is one of the world's largest vertical farms, with vegetable plants stacked from floor to ceiling inside a former steel mill. Each plant at the farm is equipped with an IoT-enabled sensor that tracks its vital statistics, including water consumption, nutrient density, and readiness for harvest. Dell Edge Gateways aggregate and analyze the data, and the system uses machine learning to optimize the growing environment. That means the vegetables' temperature, water, and light are adjusted automatically, and the team knows exactly when to pick them for maximum flavor, nutrition, and freshness, avoiding spoilage and waste.

Green Biz published an article in 2018 entitled, "Why Data is an Essential Nutrient for AeroFarms Crops" (Clang, 2018). Urban agriculture pioneer AeroFarms eschews pesticides and herbicides. It gets away with using considerably less water than traditional growers of the leafy greens in which it specializes—it squeezes out almost 95% of what's traditionally used. But there's one ingredient it can't go without: data.

That imperative drove the well-backed startup's partnership with information technology giant Dell. Two big projects are underway there, within the 70,000-square-foot facility that houses Aerofarms' ninth indoor farming operation in Newark, New Jersey.

The first initiative uses sensors to track information at virtually every step of the growing process—from seeding to germination to growing to harvesting and packaging—and send it wirelessly to servers where it is closely analyzed. Aerofarms uses that information to improve taste, texture, color, yield, and nutrition metrics for its crops, according to a case study published by the two companies.

The second project employs special cameras to track the spectral conditions of the grow trays, and of the lighting technologies crucial for nurturing arugula, kale, and mustard greens—products that AeroFarms sells to local supermarkets under the Dream Greens brand (It nurtures 400 plant varieties.) If something unusual is detected, an alert is sent to a ruggedized tablet computer. The images are also collected and analyzed.

"We have this fully connected farm that is ever becoming even more connected," AeroFarms co-founder CEO David Rosenberg told me Tuesday during the Techonomy conference in New York. "That enables us to both manage the farm as well as take information from the farm and send it to the right people to make the most of that data." AeroFarms relies heavily on real-time information for food safety and operational processes. It can produce a crop in just 15 days: It shrank that 1 day using its information metrics, but the data is also used to influence taste and texture.

"How we organize and manage that data, it's incredibly important," he said. "When you have that as your lens, in looking at a business, you see problems in different ways and solutions come and get prioritized in different ways. That's OK."

That information will be critical for automating vertical farming processes to the point where they can be commercialized more "meaningfully." One reason AeroFarms dismantled its new facility's predecessor was that it didn't have the scale to be automated effectively, Rosenberg said.

AeroFarms in-house horticulture expert, Dr. April Agee Carroll, a pioneer and leader in the field of plant phenomics, uses phenomics to transform agriculture (Aerofarms Blog, 2018).

Phenomics is the study of all observable characteristics of a living organism. These observable characteristics are called phenotypes, and are anything measurable within the plant, such as the chemistry, color, genetic code, size, and so on. I like to call it measurement science, because we are constantly gathering all kinds of data and from there, determining what is significant to the problem at hand. Dr. Carroll utilizes advanced imaging technologies, molecular analysis, and data science to model plant growth and development to understand how plants respond to their environments. Plant phenomics is particularly interesting to me because of the amazing adaptability of plants. Plants are sessile organisms, meaning they can't pick up and move. Unlike animals who can burrow or run away from danger, plants must stay in a fixed location when facing threat. Because of this challenge, plants have adapted to their environments extremely well. Understanding this adaptability and the behavior of plants is an interesting challenge.

Corn and soybeans, for example, have long growth cycles, in months. At AeroFarms leafy greens only have 12–16 days from seed to harvest so it requires working quickly and decisively to determine the appropriate hypothesis and measurement procedure.

Although the short cycle is a challenge, it does have the benefit of rapid learnings. If you think about a traditional field farmer, they may only have 1–3 harvests per year to learn from. AeroFarm is harvesting 22–30 times a year. That's 30 chances to learn from our greens and further refine our growing algorithms.

Another exciting challenge is how to manage the tremendous amounts of data gathered from each harvest! These rich data sets are the key to continued learning.

Nestle believes that the future is bright for AeroFarms and the controlled agriculture space and agriculture overall. AeroFarms is combining top practices of horticulture, engineering, and data science like never before. They are working with the Foundation for Food and Agriculture Research to optimize the plants for nutrition and taste.

FIGURE 6.6 AeroSpring gardens.

AERO SPRING GARDENS (WWW.AEROSPRINGGARDENS.COM)

Growers Vertical Systems Indoors or Greenhouses—Singapore (Young, 2018)

Aerospring Gardens is home-grown in Singapore—the dream of Thorben and Nadine Linneberg, who were keen to grow fresh produce in small spaces with no pesticides, and who found a way to turn their dream into a successful consumer product (Figure 6.6).

The Aerospring is an easy-to-assemble vertical aeroponic gardening system for indoors or outdoors that can be put together in around 15 minutes. The 75-L bucket has 9 or 12 hexagonal planter sections made from high quality, UV-stabilized, food-safe plastic; these all fit together in around 15 minutes. Each Aerospring can grow between 27 and 36 plants. These gardens can grow basil, mint, chillies, tomatoes, cucumber, sage, lavender, lettuce, eggplant, and kale!

AESSENSEGROWS (WWW.AESSENSEGROWS.COM)

Manufacturer of Vertical Stacked Systems Indoors/Cannabis—California (AEssense website)

AEssenseGrows (pronounced "eh-sence") was founded on solving three 21st century issues: the need for high growth yields for a growing population, food safety, and conservation of resources.

To attain these goals, it had to re-invent the entire process of agriculture down to its purest "essence." They created the AEtrium series, accelerated plant growth platforms and software delivering pure, zero pesticide, year round, enriched growth. This is the safest, cleanest, and most sustainable food & medicinal production available. While technology in farming has advanced considerably in the last century, they believe a new growing technology is the next revolutionary jump in agriculture. With decreasing available resources and a human population fast approaching 9 billion, that same leap is what the world desperately needs. Their products for growing fresh vegetables and leafy greens include:

- AEtrium-SmartFarm 4-Layer Vertical Stacked Aeroponic Grow Environment
- AEtrium-4 Tall Plant Aeroponic SmartFarm
- AEssense Lighting
- Guardian Grow Manager Automation Software
- AEtrium-2.1 SmartFarm For Cloning
- AEtrium-4 High Yield Blooming System

AEssenseGrows has been working in the cannabis growing segment. Here are some current press releases concerning their company which are on their website (Figure 6.7).

Rocky Mountain Marijuana Inc. Selects AEssenseGrows Aeroponic System as the Cultivation Technology for its Cannabis Business (Jan. 8, 2019)

AEssenseGrows Introduces 'the Perfect Grow Light' for Cannabis (Nov. 13, 2018)

AEssenseGrows Receives UL Listings for Complete Indoor Grow System (Nov. 13, 2018)

AEssenseGrows to Showcase Precision Aeroponics System at NCIA's Annual Cannabis Business Summit & Expo (July 10, 2018)

AEssenseGrows is located in Sunnyvale, CA and Shanghai, China.

In 2018, AEssenseGrows co-sponsored an International Indoor Farming Symposium in Shanghai on commercial indoor agriculture with the Shanghai Academy of Agricultural Sciences (SAAS) that attracted 160 attendees (Gibson, 2018). "Indoor farming is the solution to sustainable food production and water limitations for the world's increasing mega-cities," Christine Zimmermann-Loessl, chairwoman of the

FIGURE 6.7 Aessence vertical aeroponics SmartFarm.

Association for Vertical Farming, told attendees on the opening day of the symposium. "Vertical farming can address the food shortages while instituting controls that improve food safety. Existing horticulture methods are not sustainable and will not meet the needs." During her presentation, "The transforming power of vertical farming—global trends and local impacts," Zimmermann-Loessl said and that vertical farming is "attracting young people to this industry."

The symposium brought together some of the world's leading experts in commercial indoor cultivation for an exchange of ideas and information about the latest innovations, technologies, and research in the field.

Prof. Erik Runkle, Michigan State University, and Dr. Ep Heuvelink, Wageningen University, Netherlands, led the chorus of speakers as they talked about the breakthroughs in both energy efficiency, yield improvements, and precision, coming through indoor LED lighting.

Dr. Dickson Despommier, emeritus professor at Columbia University, predicted indoor farming will play a large role in the city of the future. "In 20 years, 80 percent of people will live in cities," Despommier said. "Cities and buildings will integrate natural resources: plants, water capture, energy management. A key first step is rooftop greenhouses on the way to vertical farms throughout the city."

Professor Toyoki Kozai, president of the Japan Plant Factory Association, discussed the role advanced technology will play in vertical farming. "Plant factories need artificial intelligence to double plant factory productivity over the next five years because there are so many complex topics to integrate," Kozai said.

"This is exactly the type of event needed to help ensure the indoor farming industry will meet the needs of society while presenting opportunities to the business community," said Robert Chen, president and CEO of AEssenseGrows. "The conversations and presentations I'm hearing so far give me great confidence that our burgeoning industry will do both."

The concept is a long way from displacing conventional agriculture, but it has inspired research, start-ups, and pilot projects in the United States, Europe, and Asia, along with imaginative artists' renderings. "It's an idea that's talked about a lot. It's still very much an emerging field where there are not that many commercial operations, especially on a larger scale," says Neil Mattson, associate professor at Cornell University in the Horticulture Section of the School of Integrative Plant Science.

With growing consumer demand for locally grown produce, vertical farming draws on specialists in the natural sciences, but also engineers and computer and data scientists. "We need engineers and plant scientists who are excited about this," enthuses Murat Kacira, professor in the Department of Agricultural and Biosystems Engineering at the University of Arizona. "For certain high-value crops and localities and climates, this technology might be economical now. But there is room for improving these vertical farming systems, especially the engineering, because the costs are still really high to operate these systems."

"This attracts the young generation, especially the millennials. They are interested in agriculture now because of the high-technology applications and potentials."

While Despommier envisioned 30–40-story establishments, current-design vertical farms typically occupy one or more stories of a building with up to 12 layers of short-statured plants stacked in trays on each floor. The plants are grown using hydroponics, aquaponics, and aeroponics. Hydroponics involves submerging the roots of the plant in water, while aeroponics is suspending them in air and bathing them with mist. Aquaponics gets into aquaculture, where you raise fish such as tilapia in the water, and nutrients, oxygen, and carbon dioxide are exchanged in a symbiotic relationship. Most vertical farms use some form of scissors lifts, ladders, or stairs, rolled around manually, to access the plants to maintain and harvest them.

Most of the crops grown are leafy greens such as lettuce, spinach, kale, and basil. Penny McBride, vice chair of the Association of Vertical Farming based in Munich, Germany, says "a lot of criticism has been that people are mostly growing lettuce. That's largely because the amount of time it takes to grow lettuce is shorter. So they can grow it quickly and get it out the door." But McBride says you could grow carrots and vining crops like peppers, beans, and cucumbers, along with tomatoes, strawberries, and medical herbs. In theory, corn and wheat could be grown as well as biofuel crops and plants used to make drugs. Transplanted seedlings are also developed, and the industry is interested in cannabis, mushrooms, and edible flowers.

Vertical farming offers many advantages over conventional farming, McBride contends: "It is appropriate closer to large population centers. You get more production on a smaller footprint." Indoor farming can increase crop yields 10–100 times over a similar footprint outdoors. Likewise, transportation of warehouse-grown crops is negligible since they are mostly consumed near where they are grown. Growth takes place at twice the rate of outdoor farming, and at least 70% less water is used than with conventional agriculture.

The speakers and presentation titles at conference included the following (Plant factory, 2018):

The Present & Future of the Plant Factory with Artificial Lighting
Dr. Toyoki Kozai—President of Japan Plant Factory Association
Tomorrow's City Will Be Self-sustaining
Dr. Dickson Despommier—Emeritus Professor, Columbia University
The Transforming Power of Vertical Farming—Global Trends & Local
    Impacts
Christine Zimmermann-Loessl—Chairwoman of the Association for Vertical
    Farming, Germany
Cuello's Law & Designing Vertical Farms into the Food-Water-Energy Nexus
Prof. Joel Cuello—University of Arizona
Research Needs in In-door Vertical Plant Production
Prof. Heiner Lieth—University of California-Davis
A Total Solution and Innovation for Smart Plant Factories with Artificial
    Lighting
Dr. Huafang Zhou—Vice President, AEssense Corporation
A New Generation of City-Industry Integration: Sunqiao Agriculture Park
    Urban Design Project
Julian Wei, AIA—Director, Shanghai Sasaki Associates, Inc.
Latest Developments in Dutch Greenhouse Cultivation & Vertical Farming
Dr. Ep Heuvelink—Wageningen University
Study of Optical Technology & Plant Physiology.

Prof. Jingquan Yu—Zhejiang University
Light Effects on Flowering and Morphogenesis of different Plants
Prof. Byoung Ryong Jeong—Gyeongsang National University
Shanghai Edible Fungi Factory Product Development
Prof. Jiachun Huang—Shanghai Academy of Agricultural Sciences
Lighting in Vertical Farms to Produce Crops with High-Quality Attributes
Prof. Erik Runkle—Michigan State University
Power Solutions for Plant Factories
Prof. Zhigang Xu—Nanjing Agricultural University
Demands For Plant Crops & Recommended Varieties For Plant Factories
Mr. Minkyu Lee—General Manager, Vilmorin-Mikado
The Industrialization, Exploration, and Business Model of Plant Factories
Prof. Wei Fang—Taiwan University
Plant Growth in Space & Mini Space Greenhouse
Prof. Huiqiong Zheng—Shanghai Institutes for Biological Sciences
Prof. Shirong Guo—Professor, Nanjing Agricultural University"
Study & Analysis of Artificial Lighting System in Greenhouse & Plant Factory
Dr. Jinglong Du—Chairman, Megaphoton
Indoor Plant Factory Development Strategies
Dr. Qichang Yang—President, China Plant Factory Association
Advanced Semiconductor Light Source Solutions of OSRAM Plant Lighting
Dr. Jiaping Shao—OS Sales Head, OSRAM GL Greater China

AGRICOOL (WWW.AGRICOOL.COM)

Growers Vertical Stacked Indoor Containers—Paris, France (Dillet, 2018)

Gonzague and Guillaume formed a French startup company in Paris to grow produce aeroponically, locally, without pesticides, and economically. They started growing strawberries in their home and then moved to a storage container, 33 m². They proved that they could produce tasty produce in what they called a Cooltainer. They raised about €4 million, and moved into a 1,500 m² factory in the Parisian suburb of La Courneuve, and hired a talented team of 30. After 2 years of R&D, they are now able to be 120 times more productive, without pesticides or GMOs, while using 90% less water and nutrients, and consuming only renewable energy. Most importantly, it's a method that brings taste back to our food, produced 100% locally (Figure 6.8).

In 2018, French startup Agricool is raising another $28 million round of funding (€25 million). Bpifrance, Danone Manifesto Ventures, Marbeuf Capital, Solomon Hykes and other business angels participated. Some existing investors also participated, such as daphni, XAnge, Henri Seydoux, and Kima Ventures. Their container can control the temperature, humidity, and color spectrum using LED lights to replace the sun. They can grow strawberries all year round, save water as a container is limited when it comes to space, and save on transportation having locally-produced, GMO-free, pesticide-free strawberries. Agricool plans to launch a hundred containers by 2021 in Paris and in Dubai. That's why the company is going to hire around 200 people by 2021 to support this growth rate. Eventually, Agricool also plans to expand to other fruits and vegetables.

FIGURE 6.8   Agricool aeroponic container system.

AGRIHOUSE (WWW.AGRIHOUSE.COM)

Manufacturer of Aeroponic growing equipment, Colorado USA (Agrihouse website)

AgriHouse is a leading-edge agri-biotechnology company offering registered, safe broad-spectrum biopesticides and advanced tech products that conserve valuable resources. AgriHouse was originally founded by Mr. Richard Stoner, CEO, inventor and aeroponic expert, and Dr. Ken Knutson, plant pathologist, Colorado State University. They are now AgriHouse Brands Ltd, a limited liability company, led by Lyric Turner, GM. AgriHouse's technology has been thoroughly tested and proven effective by the USDA, NASA, NSF, US Forest Service, leading universities, agricultural extension services, and worldwide corporations, farmers, and growers (Figure 6.9).

FIGURE 6.9   Agrihouse aeroponic walls.

The company offers patented & patent-pending products to deliver cost-effective ecologically beneficial solutions to boost food production and harvest yields. Our homeopathic bioactive products will help sustain the environment, conserve water and natural resources, and reduce the reliance on toxic pesticides. Agrihouse was Aeroponics International in the 1990s and EnvironGen before that. They collaborated with NASA in the 1990s. They currently offer a series of products including an aeroponic system, biocontrols, and leaf sensors.

AMPLIFIED AG (WWW.AMPLIFIEDAGINC.COM)

Manufacturers, Growers, and Sensor Designers – Vertical Stack Systems—South Carolina (Amplified Ag website).

Amplified Ag is a family of companies located in Charleston, SC with a mission to create a global shift in the production, distribution, and consumption of leafy greens. They are accomplishing this by innovating a more powerful and efficient way of farming through state-of-the-art technologies. In doing this, they are engaging a new generation of farmers and providing them with sustainable jobs. Amplified Ag consists of three companies—Tiger Corner Farms, Boxcar, and Vertical Roots. Tiger Corner Farms is an indoor aeroponic container farm manufacturer dedicated to building high quality, efficient indoor container farms in order to create opportunities for farmers. Boxcar is a technology company that provides the controls and the software needed to integrate, automate, and monitor all aspects of an enterprise farming environment, from seed to sale. Vertical Roots is a farming operation that currently services over 50 restaurants, 4 grocery store chains, and 25 local public schools using their technology. The founder and CEO is Don Taylor, a software industry veteran of more than 25 years. Don is passionate about technology and software innovation. His experience in managing large, complex technological environments across multiple industries now drives his mission to help develop a sustainable farming ecosystem while creating successful business opportunities for farmers (Figure 6.10).

CLOUDPONICS (WWW.CLOUDPONICS.COM)

Manufacturers of Controllers—US (Cloudponics website)

The Cloudponics GroControl is a controller that allows one to remotely monitor and automate ones indoor grow with nutrient dosing and pH sensors. The GroControl pumps up to three nutrients and pH up & down buffers into ones reservoir to keep the electrical conductivity (EC) and pH level within the optimal range at all times. Cloudponics sells a Hydro-Aeroponic home use system called Aero Force One. The Aero Force One Hydro-Aeroponic system is custom designed to work with the GroControl. It includes a patent-pending dual chamber reservoir that refills the nutrient mix with fresh water whenever needed.

The system is powered by a 400 gpm water pump that keeps the roots in an oxygen-rich environment, maximizing nutrient uptake, and encouraging roots to grow with vigor. The 15 gallons fresh water reservoir is kept oxygenated with an air pump. It uses an iPhone & Android app to control the system (Figure 6.11).

FIGURE 6.10   Amplified Ag/vertical roots container farm.

(*Continued*)

FIGURE 6.10 (CONTINUED)   Amplified Ag/vertical roots container farm.

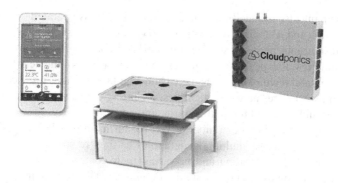

FIGURE 6.11   Cloudponic aeroponic systems and devices.

In 2018, Cloudponics, maker of the GroBox, the intelligent, fully-automated IoT aeroponic plant grow system for the home, announced that its two products, the GroBox and GroControl, are available for European customers. The GroBox and GroControl is available for sale in the following countries: Austria, Belgium, Czech Republic, Denmark, Finland, France, Germany, Greece, Ireland, Italy, Luxembourg, Netherlands, Norway, Poland, Portugal, Spain, Sweden, Switzerland, and the United Kingdom.

The GroBox intelligently monitors and manages multiple variables such as air temperature, nutrients, humidity, water flow, airflow, light schedule, and pH balance in order to create and sustain the optimal conditions for consistent, predictable, and repeatable yields. The GroBox can be paired with the Cloudponics app for 24/7/365 oversight and can run for up to three weeks before a water refill. The Cloudponics technology can be used to grow a variety of plants and produce, however, the GroBox has been designed specifically for the unique needs of cannabis growers, with special features designed to ensure the security and discretion of your plants, such as an odor filter, app controlled door lock, and lights specifically designed for growing cannabis.

80 Acres Farms (80acresfarm.com)

Growers Vertical Stacked Systems Indoors—Ohio (80acresfarm website)

80 Acres Farms, an Ohio-based vertical farming startup, has raised private equity funding from Virgo Investment Group, a private equity firm from San Francisco, for the construction of what it says will be the first fully automated indoor farm (Burwood-Taylor, 2019). *AgFudnerNews* indicated that the deal was worth more than $40 million in equity capital.

The funding will go towards the completion of 80 Acres' Hamilton, Ohio facility, which was announced in 2018 and is set to be partially operational in 2019. It will be automated from seeding to growing to harvesting featuring handling robotics, artificial intelligence, data analytics, and around-the-clock monitoring sensors and control systems to optimize every aspect of growing produce indoors (Figure 6.12).

The company also has facilities—that are more manually operated—in Arkansas, North Carolina and Alabama from which it serves local major national grocers, local retailers, restaurants, and food service companies with leafy greens, tomatoes, micro greens, and herbs. It expects to add grapes and strawberries in the near future.

The design and technology of the new Hamilton facility takes into account years of research by 80 Acres across its facilities, including comparing produce grown in greenhouses and outdoors for quality, taste, and nutrition.

While there are a growing number of vertical farms in the US and globally, few have reached scale across multiple geographies as they've struggled with the high costs associated with operating and powering their controlled environments, particularly when it comes to staffing and LED lighting. This has meant the unit economics of producing food have not always made sense. There has also been a wave of undisclosed food safety mishaps including pest outbreaks that caught some less-experienced operators by surprise.

FIGURE 6.12   80 Acres farm.

The experience of the 80 Acres team, which has worked at several large food businesses in the US including Del Monte, and its focus on food safety was a big draw for Virgo. Virgo researched several other indoor farming groups before making this investment.

"The 80 Acres' team was the only team that's run large-scale food manufacturing facilities in the past," said Eli Aheto, a partner at Virgo. "And you can see it in the design of the facility. Every element that's in contact with the plant has been thought through with a safety perspective, such as how the water comes in, is used and filtered, and all the air handling equipment too. This is all very controlled and unique compared to other folks with big open facilities."

Those big facilities have no way of preventing pests or disease from spreading throughout, while 80 Acres have designed a sectioned facility to address this, added Aheto saying he had heard about outbreaks at other facilities.

Unit economics and 80 Acres' robotic technology efforts were another key differentiator for 80 Acres encouraging Virgo to make the investment. "Vertical farms need to be able to operate facilities in a cost-effective manner and our goal is to produce a product that's affordable and available to all consumer segments, so the key is to automate as much as possible. Some other facility designs don't lend themselves to effective automation; it's one thing to design automation in a vacuum and it's another to understand automation in a food facility and reflect that in a farm design, knowing what can go wrong in a facility. 80 Acres has designed a system that can effectively automate production in a way that's still safe."

80 Acres did not decide to build all of its technology in-house as many indoor farming groups have and instead partnered with various groups on developing the technology it uses, including Priva, the well-established Dutch horticultural grow equipment group.

"When it comes to unit economics, it's about labor costs, with energy costs a close second," 80 Acres president Tisha Livingston told *AgFunderNews*. "Most indoor farms are very manual and labor intensive so being able to make that leap from manual to automated is really important to get the unit economics right."

80 Acres Farms CEO Mike Zelkind said that the secrecy surrounding the stack technology used in vertical farming is starting to mature. "Because this is such a capital-intensive industry there has been a lot of secrecy because farms have been constantly fundraising, which is not conducive to a friendly industry," he told *AgFunderNews*. "But that's changing and we're sharing more with our friends in different geographies as we all realize there's a lot to gain in working closely together to ensure this industry grows." "Particularly we think that food safety practices should be shared to protect the industry overall. There are a lot of entrepreneurs coming into this space with great background but they haven't been through a bunch of recalls and you really can't just read about those to understand the severity of them," he added.

FarmedHere

Growers Vertical Farms (Hardei, 2017)

In 2017, FarmedHere, one of the nation's largest vertical farms shut down. It was also the first to be certified organic by USDA. Rosemont, the FarmedHere founder, indicated that this isn't the first of the capital-intensive vertical farms to close. Mirai, Panasonic, and Google have all abandoned vertical farming projects. Does that mean that the money just isn't there to do these things commercially at a large scale? Basically, what the owners of FarmedHere said is that they weren't making enough money and could leverage their investments better in other sectors. At the same time there have been small-scale vertical farms with tons of success. So, is the answer that without government or other outside assistance (like Sky Greens in Singapore, which is still expanding), it's up to smaller farmers with better sales channels to develop the technology before vertical farming really becomes scalable? There are still big vertical farms that look to be doing well. It's just a matter of adapting and continuing to learn—and this can happen at the same time with both highly funded ventures and smaller models.

GREEN HYGENICS HOLDINGS (WWW.GREENHYGIENICSHOLDINGS.COM)

Manufacturer Hybrid Aeroponic Systems/Cannabis (Green Hygienics Holding website)

Green Hygienics Holdings Inc. a full-scope premium cannabis cultivation US-based company targeting the high-end medical and adult-use recreational market. With more than 25 years of experience in agricultural science and innovation, Green Hygienics is establishing itself as a leader in the advancement of science-driven cannabis cultivation systems. Green Hygienics employs scientific methodology to create a sterile growing environment by utilizing hybrid-aeroponics for a soil-less approach. This advanced hybrid aeroponic method, along with other proprietary techniques, produces the highest yielding, lowest cost product currently available within the global cannabis marketplace. Green Hygienics Holdings, Inc., has extensive expertise in indoor horticulture and vertical farming. Requiring no natural sunlight or soil, their hybrid-aeroponic systems can be in operation and producing anywhere in the world, 365 days per year. Green Hygienics' vertical farming using aeroponic technology uses 90%–95% less water, less energy, and improves yield per square foot significantly. The system mists water and nutrients directly onto the roots in a controlled environment, dramatically reducing spoilage while keeping the farm bug free and 100% organic.

GROW ANYWHERE

Grower—Colorado (Higgins, 2008)

Grow Anywhere started in 2005 in Colorado when Larry Forrest, optometrist, decided to try something new. Richard Stoner, president of AgriHouse Inc. in Berthoud, told Forrest about aeroponic plant crop systems and persuaded Forrest to start one. Forrest grows more than 12 types of crops at a time. These 2-inch tall plants are grown aeroponically—without dirt—to sell as garnishes to local restaurants.

Grow Anywhere produces the greens of beets, broccoli, basil, cilantro, arugula, celery, carrots, wheatgrass, peas, dill, fennel, and onions. "It is a pretty unusual business," Forrest said.

Grow Anywhere grows the indoor plants suspended in air on soft meshing. The growing area is 1,800 square feet inside a 3,000-square-foot industrial warehouse. The plants are kept watered every 20 min with a nutrient filled sprinkler system, and they grow in 70° temperatures under full-spectrum lights.

"It takes up to two weeks from the time the seed is planted until the plant is harvested," Forrest said. Scissors are used to cut the small plants during harvest.

"This is a very expensive product to start," Forrest said. "Rick was the brains behind this." Stoner sold Forrest the equipment and license to use AgriHouse products.

"We sold him the products, but he implemented the technology," Stoner said.

Grow Anywhere's first year of business was a learning process.

"It was a trial and error to see what would work," said Gail Stout, Forrest's sister and one of Grow Anywhere's four employees.

"I was changing our process every week before we got it down."

Mark Haberer, a former chef, is the operations manager of Grow Anywhere.

"To keep crops tasting fresh, Grow Anywhere uses no pesticides or preservatives," Forrest said.

"Our crops are only sold locally."

To appeal to specialty restaurants, Grow Anywhere also sells micro-mixes of its plant crops to create specific flavors for dishes. The base for these mixes contains mustard greens, lettuce, bianca rica endive, onions, celery radish, and carrot cuttings.

To create a Mediterranean mix, the base is added to beet for color and basil and fennel for flavor. The Atlantic mix adds dill and broccoli. The Asian mix, which adds shiso and shungiku, goes well with fish dishes, Forrest said. Other specialty mixes include a basil mix, a Latin mix, a color and spice mix, and a spring onion mix.

Boulder and Denver restaurants are the biggest buyers of these garnishments. Restaurants purchasing the plants include Bloom, Jax Fish House, Sushi Tora, Bacaro, and the Blue Sky Grill at Coors Field.

The colorful garnishes not only make a dinner plate more aesthetically pleasing, but also taste good, Forrest said.

The micro plants range from sweet to spicy in flavor.

"I don't ever garnish my plate; I eat it," said Stout, who harvests the plants. "I think they are missing the boat when they don't eat them."

"Garnishments are used to enhance color, texture and taste to any dish," Forrest said. "A restaurant may put a few onion sprouts on top of a soup to add some flavor."

Grow Anywhere ships more than 100 2-ounce clamshell containers of garnishments twice a week to local restaurants.

In an interview with Food Engineering magazine, Forrest gave the following answers to their questions.

FE: The misting process is called hydro atomization. Please describe.
Forrest: The water and nutrients build up to 100 psi, then pressure is released as they go through the hydro-atomizing jet. There is a little plate at the front of the misting orifice that fractionates each water droplet to a size of 5–50 μm.

FE: Are there maintenance issues with the nutrients clogging the jets?

Forrest: In the 1980s, we relied on a brass jet with a steel impingement arm that came over the orifice. Because you had two dissimilar metals, and because a current runs through the water, there was calcification and mineral buildup. With the help of one of our NASA grants, we were able to design a nylon nozzle that resolves the issue.

FE: How are plants supported?

Forrest: The original system was modified in partnership with Bio Serve Space Technologies with the second NASA grant. It looks like an inverted volcano, with very smooth surfaces on the inside. Separating the top of the plant from the roots is a microfilm with very low mass. The roots grow through the proprietary film, which essentially supports the plant.

FE: What types of produce are you growing?

Forrest: Cilantro, arugula, onions, beets, corn, and pea sprouts—it's truly garnish for restaurants. Everything is sold in a 2-ounce box for about $8, and each box might complement 20 plates. Restaurants that charge $16 or more per serving usually rely on Mexican suppliers who are three days away from them for these ingredients. Our microgreens are hours old when they arrive at Colorado restaurants, and they're healthier and tastier.

FE: How does aeroponics compare with hydroponics?

Forrest: The water and nutrients required are significantly less with aeroponics. In the low-gravity tests on the space station, aeroponics required 65% less water and 45% less nutrient inputs than hydroponics, while increasing the amount of biomass by 80%. Instead of immersing the plant's roots in water, they are exposed to the oxygen they need all of the time, except for the three seconds of misting that occurs every 30–40 min.

Hydroponics is used extensively for tomatoes, but it hasn't been particularly successful for other crops, partly because it is energy- and water-intensive. Aeroponics is radically different because it uses a fraction of the water and energy. Our plants are hanging in the air, not sitting in water. We only use a few gallons of water and solution a day. It's something like the misting systems in supermarkets' produce sections. The biggest expense is artificial lighting.

FE: How large a yield do you get?

Forrest: We generate 250–300 2-ounce containers a week in a growing area that's 1,800 square feet. We use scissors to harvest the crop, which sounds labor intensive but probably is less so than in conventional operations, where microgreens have to be washed and dried before shipping. These are delicate crops, and gentle handling is critical. With aeroponics, a couple of people can harvest 100 boxes in about three hours.

FE: What challenges have you faced with the greenhouse?

Forrest: We killed a lot of plants the first year due to interruptions in the misting. If the lights are on and the water isn't flowing, you can cook the plants in a couple of hours.

An unexpected problem early on was the alkalinity of the water. Plants were turning yellow, and we had no idea that water pH was even an issue.

It turned out that pH had to be in the 6.4–6.8 range. Now we balance the acidity in large containers after going through a reverse-osmosis process. Nutrients are then added. Early on, we were adjusting the pH twice a day, but now adjustments are only made occasionally.

FE: What have you done to optimize production?

Forrest: There are 15 water jets servicing each of the 60 frames supporting the plants, and additional jets for the germination racks that extend 10 by ·50 feet. Four connections have to be made for each jet, or about 4,000 in total. We can reduce that to a total of 40 jets in a moving system that goes back and forth over the farm, sort of like an automatic garage door.

A microchip was used to control the delivery of the atomized nutrient spray in his design 25 years ago. The current system's controller is the size of your hand, and now everything is solid state, no mechanical relays. But the same logarithms developed in the '80s are wired into today's system to ensure proper delivery.

FE: Would the economics improve with more intense cultivation?

Forrest: Yes. To be economically feasible, it probably should be two or three times bigger than the system we have in place. This could be accomplished in our building by piggybacking two plants. We only go up about 8 feet in a building with a 20-feet ceiling. And there are plans to build a plant the size of a football field in Dubai.

Our goal has been to put together a system that can be duplicated in another location and work as advertised. We're just about there; now it's a matter of expanding distribution through our national produce-distribution partners.

Gro-pod (www.gro-pod.co.uk)

Manufacturer and Training Aeroponics—UK (www.gro-pod.co.uk)

As soil-less growing systems like hydro- and aeroponics become more common, Dan Hewitt thinks it will be necessary for agricultural education to evolve too.

Alongside the commercial applications of its pods, the business has a vision to become a training center focused on aeroponic growing, supported by specialist courses at the region's colleges. "We are quite hooked on getting this to a place where Gropod is a training center to get the farmers of the future ready to go out on to the farm with pods," Mr Hewitt said. "A traditional farm manager is used to managing soil, which is very different. We think there will have to be investment in people as well as the technology." Mr Hewitt said Simon Coward, director of Hethel Engineering Center in the UK, which supports Scottow Enterprise Park, is also keen to make Gropod a "center of excellence" for training in aeroponics.

With pressure mounting on finite land supplies, a Norfolk firm is changing the game for potato production—speeding growth by giving growers complete environmental control inside aeroponic "pods." Enhancing the flavor of vegetables, speeding growth and optimizing water and nutrient use are some of the possible benefits from a "pioneering" technique being practiced by a Norfolk start-up. Gropod, a designer and manufacturer of aeroponic growing pods based at Scottow

FIGURE 6.13   Gro-pod growing sweet potatoes.

Enterprise Park, is involved in a trial with a major regional food producer to see if its technology can be used to grow tropical crops in East Anglia (Figure 6.13).

With this successful project, the company is already thinking of national applications for its products to help alleviate the pressures on arable land—and scientists involved in the project think it could open exciting opportunities in crop biology.

The company was founded in 2016 by agricultural sales veteran Dan Hewitt, who most recently worked as managing director of a potato supply group in north Norfolk, who calls the firm a "pioneer in aeroponic root production."

The key to its methods are insulated pods over which farmers can have total environmental control. The aeroponic growing process, which sees the roots exposed to the air, uses a high-pressure device which mists the plants with water and nutrients at regular intervals.

The water is captured and reused—Gropod estimates its growing process uses 70%–90% less water than soil growing—and is also analyzed to assess any nutrient deficiencies in the plants.

The company is currently using the technology to grow sweet potatoes in a trial for Kettle Foods, which challenged the John Innes Centre (JIC) to explore the possibility of growing such tropical crops for its products locally.

Gropod was put forward by JIC, and received an Eastern Agritech grant and a cash injection from Kettle to develop its pods.

They were, and continue to be, supported by Dr Jonathan Clarke, head of business development at JIC, who had met a UK-based sweet potato grower working in Essex and Kent had some ideas of the potential problems which Gropod's technology could try to address.

After working for a year to design the 21 m² pods, which were planted with 320 sweet potato plants each, Gropod made its first harvest in December 2018.

Another project, growing blue and red potatoes, has already finished and been deemed successful—in the pod's "optimal" conditions, the plants were producing tubers within 5 weeks and were harvested in eight.

Mr Hewitt said: "We can grow tubers to harvesting in six weeks—that's eight or nine crops a year. Take the environmental factors away, give the plant the food it needs 24 hours a day, and you have optimal conditions which will produce perfect crops."

Mr Hewitt is keen to reduce the pods' environmental footprint further by using solar power to run the pods' lights—currently high-pressure sodium bulbs, but LEDs are being trialed.

Through their process Mr Hewitt and Mr Wright also made the "surprise" discovery about the air temperature needed for aeroponic growing. "When you plant a potato in a field you want soil temperature of 9°C. When you don't have soil the air temperature around the roots is critical, and it is a lot higher than I thought it would be," Mr Hewitt said.

Dr Clarke said improvements were evident in the second generation of pod infrastructure, which was more efficient, and was helping the plants to grow more vigorously and produce better tubers more quickly.

The technology would start to "re-educate" people about crop growth, he said.

"You tend to leave it to Mother Nature, but with this you have a strict regime you can control. Suddenly the plant is growing differently and you realize you can influence it."

"Gropod has heralded a change in vegetable production strategies and I think it is going to offer producers and growers something exciting."

JIC researchers will analyze the crops produced to see how the level of control enabled by aeroponic growing environments could be taken advantage of, for example to adjust the nutritional value or flavor of the tubers or bring back heritage or heirloom varieties which do not manage well in traditional soil growing.

With land for food crops likely to become a rare commodity in the future, investors and producers are responding to the idea of pod growing. Last month potato giant McCain announced an investment in an "urban farming" company in Canada, using similar techniques to Gropod, proving demand for less resource-hungry growing solutions.

"Commercially it works, and there are a lot of people out there who want to see it succeed," said Hewitt. A major goal for Gropod is the development of "sustainable supply chains," which can reduce the air and road miles on fresh produce. Hewitt added: "Rather than importing sweet potatoes from all over the world, can we get our potato growers to put in pod infrastructure for them to carry on growing ingredient crops all through the year?"

"Using clever technology, water and solar and making use of redundant spaces—which there are thousands of—the importing of food becomes less critical."

As soil-less growing systems like hydroponics and aeroponics become more common, Dan Hewitt thinks it will be necessary for agricultural education to evolve too.

Alongside the commercial applications of its pods, the business has a vision to become a training center focused on aeroponic growing, supported by specialist courses at the region's colleges.

"We are quite hooked on getting this to a place where Gropod is a training center to get the farmers of the future ready to go out on to the farm with pods," Mr Hewitt said.

"A traditional farm manager is used to managing soil, which is very different. We think there will have to be investment in people as well as the technology."

GroPod® is a Science, Engineering, and Technology company that develops crop production systems for the agriculture, horticulture, and food processing sectors.

GroPod®, designs, fabricates, and demonstrates contained production systems to meet the specific requirements of its clients. It establishes crop production on a demonstration scale, defines the economics of production, secures production contracts within the agri-food supply chain, and then links production to growers and producers.

GroPod® provides services to three principal clients: food processors (fresh ingredients), retailers (fresh produce) and grower/producers. GroPod®'s systems provide its clients with climate independent, low-input crop production systems capable of delivering resilient, year-round, on demand production of both UK adapted and non-UK adapted crops.

The company was formed to design and build modular "pods," for vertical growing. The pods are designed for a range of applications for growing food ingredient crops for local supply chains.

The business was awarded funding through the Eastern Agri-Tech Growth Initiative to support a research and development project to prove the concept of 'Pod' production on a pilot commercial scale.

The next stage is to upscale the pilot in order to meet the needs of a local client.

Primarily the research is to enable the growing of sweet potatoes for the snack food industry. The crop is currently freighted from the USA, South America and Egypt, so enabling sweet potatoes to be grown locally, would reduce food miles, and the associated carbon footprint and overall operating costs.

GROWX (WWW.GROWX.CO)

Grower Horizontal Indoor Systems—Netherlands (Growx website)

Growx is a Dutch startup aeroponic company in Amsterdam with a motto "farming, straight up" using horizontal indoor growing systems. On its website it states that "urbanization, it's the trend that doesn't seem to be going anywhere anytime soon. Just look at the Netherlands, which currently requires an area three times bigger than its land mass just to meet the food demands of its citizens. While traditional farming can sustain rural communities, city-dwellers of the future need a fresh food infrastructure they can rely on—without exceeding their urban footprint. And that's precisely the kind of green revolution we're starting with GROWx."

GROWx claims to be the first vertical farm in the world to run off renewable energy combined with circular packaging, with minimum waste and emissions. Plus, it reduces the city's dependence on expensive imports with large carbon footprints. The farm's smart LED light system ensures every day is a Summer's day without a cloud in the sky. Working with expert chefs they guarantee rapid growing and predictable nutrient content—meaning the same high quality every single time (Figure 6.14).

FIGURE 6.14 GrowX home growing system.

GROWx advocates for clean food, i.e., their craft greens are always free from pesticides and its wholesome taste has even earned them a spot on some of Amsterdam's finest restaurant plates. On their website they highlight broccoli, fennel, mustard mizuna red streaks, and radish rioja. They claim that their skyscraper farms use minimal surface area so we can make the most of every square meter; they use 95% less water than the average field farm, and they completely off renewable energy.

HELIOPONICS (WWW.HELIPONIX.COM)

Innovative Aeroponic Manufacturer—Indiana (Helioponics website)

Scott Massey met Ivan Ball while working together on a NASA funded research project at Purdue University as undergraduate students under Dr. Cary Mitchell. What started as an introduction to the innovating technology of hydroponics, left them both wondering why there was not a standard, aeroponic device for consumers to grow their own produce (Helioponic website). At that moment, Heliponix™ (formerly Hydro Grow) was founded. What started as a dream for these young engineers who wanted to make a difference in the world became a literal kitchen startup in an apartment on campus at Purdue University as Ivan and Scott built their initial prototypes which were funded by working night shifts as full-time students.

"We chose the name "Heliponix" by combining the word "helix" with "hydroponics." Our goal is to find the most efficient form of agriculture, and that means farming with the least amount of space, energy, and water without compromising our commitment to growing the highest quality food. For inspiration we looked to nature, and were influenced by the helix found in DNA segments. Through survival of the fittest, nature has already determined this to be the most efficient shape, and this has been a powerful influence on our work."

Scott Massey, CEO, while pursuing his B.S. at Purdue University in Mechanical Engineering Technology and Certificate of Entrepreneurship and Innovation exited a career as an engineer in the oil and natural gas industry and part time patent drafting for a local patent attorney to become a research engineer at the Purdue University Horticulture College under Dr. Cary Mitchell. This inspired him to found HELIPONIX© in his senior year. He is responsible for overseeing next generation products, manufacturing operations, and the strategic plan of the entire business. The company employs several engineers and is expanding their network of GroPods deployed in the market which has been named the "Indiana's Best New Tech Product for 2018" by TechPoint through the Mira Award. Scott continues to advise the Department of State through the Mandela Washington Fellowship on several aquaponic farms across Africa to fight food insecurity in the developing world.

Ivan Ball, CTO, obtained his B.S. in Electrical and Computer Engineering Technology with a minor in Organizational Leadership and Supervision at Purdue University. Throughout his undergraduate Co-op, Ivan advanced his knowledge with focus towards automation controls in a large-scale food processing facility. The NASA research project sparked his fascination with the idea that anyone could continually grow their own produce without having to do manual labor in outdoor garden. Ivan is responsible for the electrical hardware design, software/web development, and cloud network architecture.

In 2018, Scott Massey gave a TED talk and Wabash College entitled Revolutionizing Agriculture with Household Appliances. In this talk he gave a future view of agriculture using aeroponic growing systems. He compared the shift from traditional agricultural practices to aeroponics like the transition from cutting ice chunks out of frozen lakes to use in ice boxes to the invention of the refrigerator. He highlighted all the advances of this new technology (Massey, 2018).

INDOOR FARMS OF AMERICA (WWW.INDOORFARMSAMERICA.COM)

Manufacturer Indoor Systems—Nevada (Martin, 2017)

Dave Martin, CEO of Indoor Farms of America, was convinced that vertical aeroponic equipment is clearly the wave of the future that IOFA is the market leader in terms of plant growing capacity and yields in any given space. According to Martin, "A recent visitor from Japan, considering a distributorship for that region, simply could not stop talking about how amazed he was at the amount of produce we are growing in such a small space. He went on to tell us he believes our container farm, with the substantially higher yield per square foot than anything else, can transform the market in Japan, and that is pretty nice to hear."

Indoor Farms of America spent nearly 2 years in R&D developing what now has multiple US patents awarded—truly affordable, economically viable high yield vertical aeroponic crop growing equipment. IOFA initially was focused on growing 30 different leafy green products but now other crops such as cherry tomatoes, strawberries, many smaller pepper varieties, and beans are being added to the capacity. The company has also tested growing larger plants, such as heirloom tomatoes, squash, and cucumbers.

In 2016, GrowTrucks, LLC container farm division of Indoor Farms of America—maker of unique vertical aeroponic indoor farm equipment, announced the delivery and operation of the first sold Container Farm using the latest patented technology in indoor agriculture, with a growing capacity that eclipses any other product in the marketplace.

"By combining our superior aeroponic vertical growing platform, with the latest components to operate the farm, we have delivered on our mission to bring to agriculture a truly economically viable solution, wrapped in a container," states Martin. The GrowTrucks' unique, reliable, and patented vertical aeroponic platform can grow up to 8,450 healthy plants with excellent flavor due to the patented system for mineral uptake in the plants, all in a 320 square foot container, and is designed to operate in any climate in the world, all year long, month after month, with no season to stop the farmer from producing. In 2017, Indoor Farms of America introduced the world's first solar powered vertical aeroponic farm. "This farm represents a major milestone for indoor farming. Everyone in the industry knows that the additional investment of solar energy generation to power an indoor farm can turn a solid return on that investment into a long term money loser, due to the length of time to recapture those extra investment dollars," states David Martin, CEO of Indoor Farms of America. "Using solar technology to power up our vertical aeroponics, which grows fresh crops that are simply beyond organic, is really a great story, if the economics work," says Martin. "Our equipment was specifically designed to create that tangible R.O.I., which is the only yardstick that frankly matters in an indoor space, and furthers how we transcend anything else in the industry." IFOA has sold containerized farm for an extreme weather area of the world, in Northern Canada, where they delivered a unit to the town of Yellowknife," says Martin. "This farm will experience temps of minus 40 Celsius, and we have spent the past year in continual R&D to ensure our turn-key farms can operate anywhere on the planet."

The area of West Africa, known as the Ivory Coast, is receiving its first world class indoor farm. "Here again, we were chosen over all competitors due to the fact our containerized farm models grow over double the yield of anything else in the world," according to Martin.

The U.A.E. is receiving its first vertical aeroponic farm from IFOA, located in Dubai, which is set to open for business in January, 2018. "Dubai will now have unprecedented access to daily fresh premiums herbs, and fresh strawberries," states Martin. Indoor Farms of America has a showroom with demonstration farms operating in Las Vegas, Nevada, and in multiple locations in Canada, and in South Africa.

Indoor Farms of America announced in 2017 that the company was launching a version of its Container Farms that will be energy-independent and water-independent, allowing deployment of the farms anywhere in the world. "We have been working on this for quite some time, and these enhancements to our farm products will be available in production form by the end of Q1 this year," according to David Martin, CEO of Indoor Farms of America.

"From the beginning of our company's existence, we have been, and remain on a path of improvement of what we know to be the best indoor agricultural equipment in existence, and to bring it to every corner of the world in a cost effective manner. We are achieving milestones each month in that regard." states Martin.

The company has container farms in operation across the US and are shipping multiple container farms to countries around the globe. "We are continually humbled by the visits we receive to our facility in Las Vegas from the brightest people in agriculture. It is what drives us to keep making it better, to fulfill our purpose as a company," commented Ron Evans, President of IFOA.

Sales and delivery of larger vertical indoor farm formats produced by IFOA, such as those in warehouse facilities, has increased rapidly, and the company is working with some of the largest companies in the world that produce or process food due to the extremely attractive financial metrics the equipment represents, according to Martin. He added: "We recently installed a warehouse farm in Salamanca, New York, when outside temps were 4°F, and the owners are amazed at what their new farm represents. They have compared our equipment to every other form of indoor growing, and chose us hands down. They are particularly focused on operating a very green farm, and use geothermal for heating of their facility, which means superior fresh greens of any kind, and strawberries, which they are starting to grow all year long, no matter what the weather is like outside."

The company has patented vertical aeroponic equipment, which provides growing capacities in excess of 40 plants per square foot of floor space in an operating height of just 8 feet.

It was announced on November 26, 2018 that Indoor Farms of America had gone out of business (igrow, 2019).

INDOOR HARVEST CORPORATION (WWW.INDOORHARVESTCORP.COM)

Manufacturer of Aeroponic Systems—Texas (Sykes, 2014)

In 2015, Indoor Harvest Corporation (brand name Indoor Harvest) is a designer, builder, contractor, developer, marketer, and direct-seller of commercial-grade aeroponic and hydroponic fixtures and supporting systems for use in urban Controlled Environment Agriculture and Building Integrated Agriculture. It is in the process of developing their products and services. Consequently, they had not generated revenues in 2015. They have an accumulated deficit and have incurred operating losses since inception and expect losses to continue during 2015. The 1Q15 and 1Q14 IHC had a net loss of $283,657 and $144,146, respectively. The increase in our operating expenses is due to increases in costs related to research and development, increased payroll costs, depreciation expense, listing expenses to have our common stock traded on the OTCQB and professional expenses related to being a SEC reporting Company. The increase in research and development expenses was due to increased costs developing our vertical farm framing system and improvements to our existing aeroponic designs.

On March 30, 2015, the Company signed a LOI with the City of Pasadena, Texas, regarding the development of a vertical farming complex, to include a for-profit commercial operation and a non-profit educational facility. The for-profit portion would include a state-of-the-art indoor vertical farm serving local customers with gourmet leafy greens, herbs, and micro-greens. The non-profit side would integrate an academic program that would use the vertical farm as a lab for practical studies and some component (to be negotiated) of low-cost or at-cost produce made available to community organizations to be balanced by tax offsets through non-profit status.

The Company was preparing to begin sales of its vertical farm framing system. Chad Sykes, Chief Executive Officer of the Company, stated, "We're taking the final steps needed to bring some of our products and services to market and based on current inquiries, we are very optimistic that we will begin sales before 2016. Current pricing of our LED vertical farm frame sets are expected to come in at a price that will compete well with most popular LED fixtures. By incorporating LEDs into the actual frame design itself, instead of using a separate light fixture and framing system, we will be able to offer a complete vertical farm racking solution for a price that is only slightly more than a standard LED fixture of equivalent wattage."

In addition to our LED framing design, we have begun the process of re-designing our aeroponic system to take advantage of cheaper manufacturing processes. We anticipate initially offering our aeroponic system as a large-scale cloning platform for propagation. We have redesigned the system to work with our existing LED frame set, thereby providing a modular vertical platform for large-scale clone production. We anticipate each unit being capable of producing some 600 clones per unit.

"Almost all current cloning systems on the market today utilize a low pressure design. By using our proven high pressure design, we expect to be the first Company to offer a true aeroponic system for commercial cloning at a price that will compete directly with less effective, low pressure systems currently on the market. With high pressure aeroponics, we are able to atomize the nutrient solution to a 30–50 μm size, thereby accelerating the rooting of cuttings several days faster than a low pressure design. We anticipate bringing this product to market in the next 2–3 months (2015) thus opening up another potential revenue stream from our existing portfolio of designs," further stated Chad Sykes, CEO.

In 2015, Tweed Marijuana Inc. announced the installation of a state-of-the-art aeroponic system developed by Texas-based Indoor Harvest Corp. (Indoor Harvest) has commenced in Smiths Falls, Ontario. The system is the basis of a Cannabis Production Pilot Agreement between the two companies. Cannabis-specific modifications will be implemented, tested, and recorded. The companies will jointly own any resulting intellectual property, with Tweed obtaining exclusive licensing rights within Canada and all jurisdictions excluding the United States.

"The Tweed R&D team reviewed a number of potential aeroponic partners from around the globe and chose Indoor Harvest for their high pressure, large-scale aeroponic specialization, and their leading edge involvement in this emerging sector," said Bruce Linton, CEO and Co-founder of Tweed. "Pushing boundaries is a part of Tweed's culture so the fit is ideal."

In 2018, Indoor Harvest has been focused on becoming a precision agriculture technology company for delivering pharmaceutical grade cannabis for researchers, and the development of next generation personalized medicines.

The CEO, Dan Weadock, wrote in 2018 that it was his great pleasure to assume leadership of Indoor Harvest Corp and to prepare the company for short- and long-term success in the rapidly evolving legal cannabis industry. The potential "legalization" of cannabis (which includes a substantial number of political, social, and scientific developments) has created what they believe are unprecedented opportunities. Indoor Harvest is seeking to take advantage of the latest technological and scientific developments.

Indoor Harvest was founded as a precision agriculture technology platform, focused on high pressure aeroponics, targeting the nascent vertical farming/controlled environment agriculture space. The original mission of the company was to revolutionize farming by eliminating harmful chemicals, reducing water and other overall inputs, while simultaneously increasing nutrient uptake, biomass, and quality to feed the ever expanding population in a sustainable way.

In 2014, in recognition of the burgeoning cannabis industry and the value it potentially represented to the company, it pivoted and turned their focus away from traditional leafy greens, toward the production of pharmaceutical-grade cannabis for the research and development community, and the exploration of new, personalized medicines. Now aeroponic platform technology is an integral element in the future of indoor farming.

They believe further that cannabis is presently the most attractive and interesting application of Indoor Harvest's technology, not only because they believe that it is a high value crop in a fast growing industry, but also because of its disruptive potential and role as next generation medicine.

Within this unprecedented market, it is anticipated that Indoor Harvest's innovative intellectual property can shine, as they refine our focus of seeking to deliver true pharmaceutical-grade product.

A new strategic partner is Leslie Bocskor with Electrum Partners, a multifaceted cannabis consultancy. Leslie agreed to work with Indoor Harvest personally and directly, to assist with strategic guidance and develop additional complementary partnerships to integrate Indoor Harvest's intellectual property into some of what we believe to be the industry's most important and valuable enterprises, seeking to ensure long-term profitability and value for the company and its shareholders.

Chad Sykes, founder and original CEO, served as Chief of Cultivation, Principal Financial Officer, and Principal Accounting Officer. It is expected that he will continue working with Indoor Harvest and will focus his short-term efforts on the development of our planned Integrated & Controlled High-Pressure Aeroponic platform showcase.

James E. Wagner (www.jwc.ca)

Manufacturer—Canada (JWC website)

James E Wagner (JWC) is a company that is revolutionizing the cannabis industry with its aeroponic cultivation technology. JWC's innovative GrowthStorm™ system, based on NASA-developed technology, allows for high yield and efficient production of clean and consistent cannabis crops. In an increasingly crowded Canadian cannabis market, the proprietary system sets JWC apart from its competitors and provides unique value to the company.

The GrowthStorm™ platform is a four-part aeroponics growing system that allows the roots of the cannabis plants to be suspended in the air in a closed-loop growth chamber. The roots are misted with nutrient-rich water at prescribed intervals, and because the roots are dangling in nothing but air, they are better able to absorb the nutrients and oxygen, which, in turn, has several benefits, including larger yields, a short growth cycle, and overall improved plant health. GrowthStorm™ uses less water and much less space to grow the same number of plants as a soil-based grow system. GrowthStorm™ also limits crop disease by eliminating the need for soil or other growing media, which are often the vehicles for disease transmission. Additionally, if a plant does become infected it can be easily removed without disturbance to the surrounding plants (Figure 6.15).

Typically grown cannabis plants have a higher cannabinoid count in the buds near the top of the plant and a lower cannabinoid count on the lower branches. The GrowthStorm™ system allows for more standardized cannabinoids over the entire plant resulting in a more consistent dose and a higher quality product for the company's patients.

FIGURE 6.15   JWC cannabis farm.

(*Continued*)

FIGURE 6.15 (CONTINUED)   JWC cannabis farm.

Additionally, the facility utilizes a series of grow rooms to enable perpetual harvesting. This allows for significantly greater yields and more accuracy in the prediction of harvest volume and timelines to meet patient demand for its products. JWC has partnered with Price Industries to create a system that results in the perfect humidity and temperature control to create an ideal growing environment, resulting in a product without any need for irradiation or cold pasteurization.

It is important to note that JWC is a family business, built on the legacy of the family patriarch James E Wagner. James was a farmer who spent most of his life out in the fields, growing everything imaginable on the family farm. Throughout his long farming career, he passed his knowledge and passion for growing quality crops on to his children and grandchildren. He grew cannabis too, and it was Grandpa Jim who offered support and advice when the first members of their group proposed the addition of medical cannabis to their business model. James E. Wagner passed away peacefully at the age of 89, in November 2017, but his legacy lives on in the family business.

JWC is looking to the future with the construction of a second site facility at its Kitchener, Ontario operation. This expansion will increase its current 15,000 square feet facility to 345,000 square feet, and it is anticipated that targeted production will increase from around 1,500 kg/year to more than 28,500 kg/year. The new facility also brings an increase in staff as well—from the nearly 50 currently, to over 400 employees. JWC anticipates that when completed, the new facility will be the largest aeroponic grow operation for any crop in the world.

In 2018 JWC announced a consulting agreement with the leading investor relations firm, First Canadian Capital Corp. ("FCC"), pursuant to which FCC will provide JWC with consulting, investor relations, and strategic corporate communications services.

JWC has a partnership with Canopy Rivers Corporation, the strategic investment arm of Canopy Growth Corporation, and boasts of a proprietary growing methodology—GrowthStorm™. Having a process that is difficult to replicate is a game changer to this lucrative industry where many of the methods used do not sustain superior plant health, consistent high quality, and high potency.

Nathan Woodworth, Chief Executive Officer and President of JWC, commented: "With the initiation of an ongoing relationship in investor relations with First Canadian Capital, JWC is pleased to anticipate an increase in our visibility. This is a key moment in cannabis history, and we are proud to be a part of it. We are happy to be working with First Canadian Capital, who we believe are the best candidate to help spread our message of the important and cutting edge work we are doing in this industry."

The Company finished 2018 with just over 190 kg of dried cannabis in its storage vault. JWC received its sales license for dried flower during 2018 and began to sell directly to patients during Q3. The first shipment to the Canopy Growth Corporation's CraftGrow store took place during Q4 and was made available to patients in early Q1 2019. The JWC customer care team focused on building relationships with several clinics and continued adding patients to the JWC roster. Over the first 2 months of fiscal 2019, the Company has seen its patient roster grow significantly, now with over 218 registered patients. The Company has also continued to ship dried cannabis to the Canopy Growth Corporation's CraftGrow store and will provide its first shipment of resins before the end of 2018, for processing into oils. The Company also expects to receive its sales license for oils in early 2019, and will make available several oil extract products for direct selling to its registered patients.

JUST GREEN

Growers Aeroponic Systems—New York (aerofarms website)

Just Greens, LLC, is doing business as AeroFarms LLC, growing leafy greens, herb varieties, and micro greens by using aeroponic technology and LED lighting. It also grows edible flowers, warm season greens, and grasses. The company serves schools and food banks. Just Greens, LLC was formerly known as Aero Farm Systems, LLC. The company was founded in 2004 and is based in Ithaca, New York.

Just Greens LLC has submitted a patent application for aeroponic system and method. This invention was developed by Edward Harwood and David Rosenberg. Just Greens LLC has sought a patent for improvement of an aeroponic system and method. This invention was developed by Edward Harwood.

LETTUCE ABOUND FARMS (WWW.LETTUCEABOUND.COM)

Grower Vertical Walls—Minnesota (lettuceabound website)

In 2018, Lettuce Abound Farms in New London, MN was started by Ortenblad and Dengler, along with their wives Julie Ortenblad and Melody Dengler, to build the region's first commercial-scale aeroponics lettuce farm.

Lettuce Abound is the first franchise operation of the Faribault-based Living Greens Farm, which developed this type of aeroponics system. Ortenblad and Dengler modified the system and, despite some challenges that delayed the start-up, they said the system is working flawlessly and plants are growing faster than expected. The facility is housed in a farm machine shed they turned into an immaculate indoor farm that meets federal vegetable production and packaging standards.

After suiting up in a white jacket and hairnet, washing one's hands and stepping in a shallow shoe wash before entering the indoor farm, visitors see vertical walls of lush green lettuce. When the facility is eventually in full year-round production, the 102- by 50-foot building will produce the equivalent of a 180-acre farm. After the seeds germinate and plants are a couple inches high, the chunks of rock wool are placed into panels with cup-like plastic containers and attached to 12 stainless steel A-frame units that are 8 feet high and 32 feet long. Each unit holds 1,536 lettuce plants.

The roots of the plants dangle in the air underneath the frame and are spritzed every 12 minutes with a mist of reverse osmosis water and fertilizer delivered through mechanical sprayers run by a complex computer program. Banks of intense lights—which are so strong that sunglasses are needed to work among the plants—help fuel the plants' rapid growth. The carbon dioxide level is carefully measured, the temperature is kept at 70–72, the humidity is at 55%–65% and a fan blows air to mimic the wind and makes the indoor farm "smell amazing," Dengler said.

During peak capacity, they will harvest three units, or about 4,600 heads of lettuce, every week, Ortenblad said. They are gradually increasing the number of varieties of lettuce and will grow specific types of greens requested by large-scale customers, Dengler said.

While they're growing just lettuce now, they have already laid the groundwork for a potential expansion in a couple years for a new building to grow aeroponics strawberries and small cherry tomatoes.

What had been a large, empty pole barn on a farm in rural New London is now becoming a climate-controlled maze of computer-operated, high-powered lights, and a water misting system housed on a dozen massive stainless steel frames. The building will be filled with lush, fresh lettuce, and other leafy greens.

It will be the region's first large-scale, commercial indoor aeroponics growing system that will produce fresh greens year-round.

They intend to market the high-end produce to restaurants and other outlets, like food co-ops.

"This is not farming. This is science," Ortenblad said. "This is science and computer automation is what this is."

LIVING GREENS (WWW.THELIVINGGREENS.COM)

Grower Indoors—Minnesota (thelivinggreens website)

In 2018, the founder of a growing aeroponic farm near Faribault, MN, CEO Dana Anderson was ready to prove his Living Greens low-input, no-dirt operation will have a big effect on the emerging world of year-round indoor vegetable growing. Anderson started tinkering with aeroponics in his Prior Lake garage in 2010. In 2017, Living Greens, staked by $8 million over several years contributed by founding shareholders, started slowly by testing and eventually producing increasingly larger crops of lettuce with a high-tech, rapid-growth system rooted in nutrient-rich misting and

LED lighting. In 2019 following completion of the last stage of construction, Living Greens should amount to 60,000 square feet of stacked, mechanized growing space capable of producing up to 3 million heads of high-quality lettuce. The business plan and initial production proved impressive enough to recently draw $12 million in an inaugural round of institutional funding from Boston-based private-equity funds NXT and Wave Equity Partners.

"We're exiting the research-and-development stage and going to market," Anderson said.

The $6 million Faribault factory-farm will prove Living Greens' technology innovation and its year-round, premium-lettuce model and spur construction of a second plant within a year outside of Minnesota, backers said.

"We think we have an opportunity to be a market leader in leafy greens," Anderson said. "There are projections nationally that up to 50 percent of leafy greens could be grown indoors within 10 years from almost nothing today." "Our goal is to be the largest indoor [farm operation]; corporate-owned and through licensing of the technology," Anderson said. "We're looking at joint ventures with food-service companies around the world."

Living Greens said that its Faribault farm—a floor footprint of about 20,000 square feet that rises to about 16 feet, thus providing its 60,000 square feet of growing space—It will be the second-largest aeroponics operation behind that of industry leader AeroFarms of Newark, N.J. It operates a 70,000-square-foot indoor farm in addition to smaller installations.

Living Greens already supplies a growing list of Minnesota and Wisconsin grocers with several types of leaf lettuce, arugula, and mixed greens through Robinson Fresh, a division of C.H. Robinson.

In 2019, it will have installed the technology to produce around 3 million bagged packages of salad greens for retail distributions.

Typical salad bags, depending on whether they include dressing and other ingredients as part of planned "salad kits," will retail for $2.99–$3.99.

"We're producing about 500,000 units a year now, and it will be a sixfold expansion by spring," Anderson projected.

The 20-employee Living Greens operation includes chief technology officer Dave Augustine, a University of Minnesota-trained electrical engineer and veteran industrialist who joined the board several years ago.

Living Greens has been awarded four patents on its growing process, according to Anderson.

The firm plans to produce as much lettuce indoors as would be produced by 100 times the farmland, using 95% less water, no pesticides, herbicides, or other chemicals.

"The risks of foodborne illness are extremely muted compared to the traditional food chain. We use a 'reverse-osmosis' process to remove all the particulates from the water."

Living Greens plans to deliver a premium product locally, at lower cost, because it won't have the long-haul transportation expense and up to two weeks' time to deliver from Mexico or the California-Texas Sun Belt where most fall-winter vegetables are grown.

To be sure, Living Greens has dreams of being a big company in a fragmented but fast-developing indoor agriculture market.

The sector is driven by the premise that farmland is limited, and industrial-scale farming can be very expensive and uses what can be unsustainable amounts of water, chemicals, and land.

The challenge for the small, indoor innovators such as Living Greens is to entice buyers with tasty produce at a competitive but profitable price that customers also will patronize for their regional and environmental pitches

Plenty (www.plenty.ag)

Grower—California (plenty website)

Plenty was founded in 2013 with the mission to bring local produce to people and communities everywhere by growing the freshest, best-tasting fruits and vegetables, while using 1% of the water, less than 1% of the land, and none of the pesticides, synthetic fertilizers, or GMOs of conventional agriculture. They use indoor vertical farms and combine the best in American agriculture and crop science with machine learning, IoT, big data, sophisticated environmental controls, and the exceptional flavor and nutritional profiles of heirloom seed stock. They are based in San Francisco and have received funding from leading investors including Eric Schmidt's Innovation Endeavors, Bezos Expeditions, DCM, Data Collective, Finistere, and WTI.

In 2017, Plenty acquired Bright Agrotech, the leader in vertical farming production system technology. Bright has partnered with small farmers for over 7 years to start and grow indoor farms, providing high-tech growing systems and controls, workflow design, education, and software. "We're excited to join Plenty on their mission to bring the same exceptional quality local produce to families and communities around the world," said Nate Storey, founder of Bright Agrotech. "The need for local produce and healthy food that fits in everyone's budget is not one that small farmers alone can satisfy, and I'm glad that with Plenty, we can all work toward bringing people everywhere the freshest, pesticide-free food."

Plenty employs 100 people in three facilities in San Francisco and Wyoming. It has raised more than $226 million since its start. Mr Barnard, the CEO, said the firm has already signed up online and bricks-and-mortar distributors for the food, which will initially be dominated by leafy vegetables and herbs (Figure 6.16).

The firm is planning indoor farms on land of 2–5 acres—roughly the size of Home Depots or Walmarts. Mr Barnard said the food will be competitively priced, thanks in part to a shorter supply chain, and within reach of a range of incomes. Tesla Inc.'s former director of battery technology joined Plenty in 2017 to lead the vertical farming startup's plan to build indoor growing rooms around the world. Kurt Kelty, who joined Tesla in 2006 and was one of the longest-serving executives at the carmaker led by Elon Musk, is the senior vice president of operations and market development. Kelty had previously spent more than 14 years at Panasonic Corp.

Plenty wants to build a giant indoor farm next to every major city in the world.

In 2017, it announced plans to build a 100,000 square foot vertical-farming warehouse in Kent, Washington, just outside of Seattle. That farm was expected

FIGURE 6.16 Plenty aeroponic system.

to be open and deliver produce locally by mid-2018, and was designed to produce 4.5 million pounds of greens annually.

Plenty grows plants on 20-foot vertical towers instead of the stacks of horizontal shelves used by most other vertical-farming companies. Plants jut horizontally from the towers, growing out of a substrate made primarily of recycled plastic bottles (there's no soil involved). Water and nutrients are fed in from the top of the tower and dispersed by gravity (rather than pumps, which saves money). All water, including from condensation, is collected and recycled.

The plants receive no sunlight, just light from hanging LED lamps. There are thousands of infrared cameras and sensors covering everything, taking fine measurements of temperature, moisture, and plant growth; the data is used by agronomists and artificial intelligence nerds to fine-tune the system. The towers are so close together that the effect is a giant wall of plants.

Currently, Plenty is focusing on leafy greens and herbs—varieties of lettuce, kale, mustard greens, basil, etc.—but it says it can use the system to grow anything except root vegetables and tree fruits. Strawberries and cucumbers are coming up next. There are virtually no pests in a controlled indoor environment, so Plenty doesn't have to use any pesticides or herbicides; it gets by with a few ladybugs. The produce from Plenty's San Francisco warehouse is certified organic, but leaders in the industry also like to stress that vertical farming is local, with an entirely transparent supply chain. Relative to conventional agriculture, Plenty says that it can get as much as 350 times the produce out of a given acre of land, using 1% as much water. "It is the most efficient [form of agriculture] in terms of the amount of productive capacity per dollar spent," Barnard has said. "Period."

The next grandest claim in the industry is AeroFarms, a Newark, New Jersey company with nine indoor farms, which says it can get to 130 times the amount of produce per acre.

What's more, Plenty says its products taste better than most of what customers now have access to. Around 35% of fruits and vegetables eaten in the US today are imported. Leafy greens travel an average of 2,000 miles to reach your plate. Some produce has been on ships and trucks for 2 weeks before it reaches the table—having lost, by some estimates, 45% of its nutritional value along the way. Produce is bred to survive that long journey with its aesthetics, but not necessarily its flavor, intact.

Plenty plans to build warehouses not inside major cities, but just outside them, next to distribution centers, to minimize the time its food spends in transit—it wants produce to go from harvest to table in hours, rather than days. If it can do that, the company will be able to grow and sell a wide variety of rare and heirloom breeds, which are more tender and flavorful than what's available at the supermarket, but less resilient to long journeys.

In fact, Barnard says he will save more money on trucks and fuel than he spends on facilities and power.

The company's goal is to build an indoor farm outside of every city in the world of more than 1 million residents—around 500 in all. It claims it can build a farm in 30 days and pay investors back in 3–5 years (versus 20–40 for traditional farms). With scale, it says, it can get costs down to competitive with traditional produce (for a presumably more desirable product that could command a price premium).

"We can pretty much grow everything," he says. "The problem is cost. Anyone can buy some shelves, some lights, irrigation. The challenge is to get your produce down from $40 per pound to $1."

His strategy for reducing costs (and improving taste) is to add data and machine learning to the traditional hydroponic mix. Throughout the farm, arrays of infra-red sensors monitor how the crops are growing and feed that information back into algorithms that adjust light, heat, and water flow accordingly.

"The best-tasting crops are finicky," he says. "They need one thing on day one and something else on day seven." The result is a closed system that controls itself, though the crops are, so far, picked by humans. "We couldn't have done this, say, ten years ago. But now we are having what I like to call a 'Google moment'. Just like Google benefited from the simultaneous combination of improved technology, better algorithms and masses of data, we are seeing the same."

"Today, with field-grown produce, 30 to 45 per cent of the final value at shelf is attributable to trucks and warehouses," he says. By cutting this journey to almost zero, Barnard says he can reduce retail prices and increase product shelf-life as pro-duce goes straight from farm to shop.

Plenty is not alone in the indoor-farming sector. There are dozens of similar companies, including London-based Growing Underground and Aerofarms in the US, that use LEDs instead of sunlight and produce fresh greens indoors. And there have been casualties. Atlanta's PodPonics, LocalGarden in Vancouver, and Chicago-based FarmedHere, all of which began with a similar vision to Plenty's, have closed down, unable to make the businesses viable. In 2019, it will open its first full-scale farm on a 9,000 m² site just south of Seattle and it has plans to move into China where Barnard sees a big market for produce that is fresh and safe.

"In China, the pesticide load is two times that of the rest of the world," he says, "so many consumers don't have the opportunity to eat fresh produce. They have to boil their veggies to feel safe eating them."

"It's a useful sideline but it's not going to solve world hunger," says Tim Lang, professor of food policy at City University in London. "I can show you books from the 1950s where people were saying that the future of food is hydroponics but it hasn't happened. It's simply very expensive to run." Long-time Tesla engineer Nick Kalayjian is leaving the carmaker after nearly 12 years to help design highly efficient vertical farms for San Francisco startup Plenty. Nick Kalayjian is a high-ranking Tesla engineer who joined in the company's shambolic early days before Elon Musk was CEO and has worked on all its products from the Roadster through the Model 3, before joining Plenty in 2018. Kalayjian became San Francisco-based Plenty's senior Vice President of Engineering. He's leaving Tesla not because of dissatisfaction or friction but out of a desire to tackle another big societal challenge after working to advance clean vehicles and energy for nearly 12 years.

"The mission of Plenty is potentially bigger than the electrification of transportation or the grid," he said. "The energy consumed, the impact on the environment of farming all over the world, is massive, and it's a problem that needs a kind of engineering focus Tesla applies to vehicles. I think there's a lot of opportunity to have a big impact." Kalayjian is a Stanford University-trained engineer. "It's super important for us to have a production system to grow food at prices that fit into everyone's budget," he said. "Few people in the world that have led engineering development and productization processes that move as fast for systems that are as dense and complex as what Nick and his teams at Tesla have done." From the outside, Plenty farms look like warehouses or big box stores, and range in size from 100,000 to 250,000 square feet, or the equivalent of 2.3 to more than 5 acres. Inside, plants grow sideways on vertical columns between 7 and 20 feet high. With precise use of water, nutrients, temperature control and light, its farms "can grow anywhere from a couple hundred acres to 1,000 acres of conventional field production on a volume basis over the course of a year," Barnard said. And they do that using 1% of the water a traditional farm would. Growing close to where the crops are consumed also cuts shipping costs, fuel use, and carbon exhaust.

PODPLANTS (WWW.PODPLANTS.COM)

Manufacturer Australian (Firestone, 2015)

Chris Wilkins was so passionate about his invention—a novel aeroponic system to grow plants by suspending them in nutrient-laden mist—that he dropped out of university to develop it.

Now the Sydney inventor is about to launch the system, PodPlants, as a means of growing greenwalls, or vertical gardens, in office blocks. But he sees this commercialization phase as only a milestone in the realization of his dream of adapting the technology to environmentally sustainable agriculture, amid fears about global food security.

Wilkins has made the finals of the Backyard Innovation category of *The Australian* Innovation Challenge awards with PodPlants. The awards are run by *The Australian* in association with Shell with the support of the federal Department of Industry. The Backyard Innovation category is open to the public and carries a $10,000 prize.

Aeroponics involves growing plants without soil. In contrast to hydroponics, in which plants are grown with their roots in a nutrient solution, aeroponics plants are suspended in air with a nutrient solution delivered to their roots in droplets, Wilkins says.

The PodPlants units are black plastic containers 2.4 m high, about 1 m wide, and 280 mm deep. About 200 plants are suspended at various levels within the modular units and are supported from their stems by foam collars fixed to the inside walls of the structures. Their leaves venture out of holes in the units' walls.

Other aeroponics systems are on the market but the novelty of PodPlants lies in part in the way the droplets are delivered to roots from a water reservoir at the bottom of the unit, Wilkins says.

Unlike conventional aeroponic systems which typically use pumps, PodPlants adapts a medical device to saturate the air with droplets without the use of pumps, nozzles or filters, he says. He has a provisional patent on PodPlants, which he has optimized for water and energy efficiency.

"The PodPlants are exceptionally efficient—consuming less than four watts of power per 100 plants and less than 400 milliliters of water per day," Wilkins says.

The portable, self-contained units, which weigh less than 10 kg, can each be installed in less than an hour, he says. The units are positioned side by side to form a greenwall.

"Traditional greenwall systems typically require plumbing, drainage and waterproofing to be installed into existing walls, making these systems impractical for many locations and costly," he says. "All the water in PodPlants is contained inside the units and the units can be placed anywhere. As there is no waste water run-off, refilling and servicing can be done fortnightly or monthly with a watering can."

Wilkins' company will initially lease the units, most of which grow rainforest plants, to businesses for installation in offices. "Eventually, we will sell them outright to plant rental companies," he says.

A former plumber, Wilkins initially spent several years working on PodPlants part-time in his backyard. He invested tens of thousands of dollars in the project (Firestone, 2015).

He was enrolled in a Bachelor of Liberal Arts and Science at the University of Sydney but dropped out to found a company and work on the project full-time.

He had participated in the University of Sydney Union's Incubate start-up accelerator program and was later invited to join the portfolio of business incubator, ATP Innovations, when he decided to take the plunge. As part of the program, he has office space at Australian Technology Park, in Sydney's inner city, and access to leading business advisers.

He was awarded a $15,000 grant by the NSW government earlier this year and tutors at the University of Sydney, and takes on interns from the institution.

Although he sees decorative greenwalls for the corporate sector as the first market for the PodPlants, he is experimenting with agricultural plants such as spinach.

"I've got two units with edibles in them," he says. "The challenge with edibles is the varying nutrient requirement."

He has experimental units in his office and home, and has even colonized his parents' and sister's homes. "Everywhere I go, I've got these units," he says.

Wilkins spends long hours working on the project and is preparing to approach prospective investors. "It's all-consuming," he says "It's all or nothing because I believe in it."

RIVIERA CREEK (WWW.RIVIERACREEK.COM)

Grower Ohio (riviera creek website)

Riviera Creek Medical Marijuana multimillion dollar facility in Youngstown, OH was built in 2018 by Kessler Products, Brian and Daniel Kessler. The facility includes offices, labs, and grow rooms. The aeroponic-grown cannabis plants are suspended, their stems and leaves virtually touching nothing as nutrients are applied directly to their free-hanging roots through a mist. The temperature, humidity, and airflow in the rooms are monitored day and night while cameras provide a means for the staff to check in on the crop from outside the room. With their new venture, the Kesslers hope to eventually deliver a medical marijuana product with the same consistency in application as any common over-the-counter drug, such as Tylenol. Riviera Creek makes up one portion of the medical marijuana supply chain made legal under Ohio House Bill 523. Under state law, dispensaries must be separate from testing labs, which must be separate from cultivators such as Riviera Creek. The marijuana business in Ohio is still very much in its infancy, to the point where all that exists currently are buildings. Testing facilities have yet to begin their work, which means dispensaries are months away from having any actual product to sell. Riviera has just under a dozen employees between its administrators, lab staff, and security personnel, though Brian Kessler expects as demand grows and the business expands, the staff will eventually grow closer to 100 (Figure 6.17).

FIGURE 6.17 Riveira creek cannabis farm.

FIGURE 6.18   Treeco tree-shaping system.

TREEVO (WWW.TREEVO.CO)

Manufacturer Israel (treevo website)

Treevo is an Israeli based company has developed a unique way of growing trees into different and creative shapes. This new product was developed to assist in tree-shaping from home. After years of research and tests, Treevo has managed to grow roots into designable molds using large-scale aeroponics techniques, a form of arborsculpture.

Treevo flora-scientists seed and sprout the plants in a specially constructed greenhouse, then elongate the roots of the tree to give the consumer a kit specially prepared for home growers. All the user has to do when they receive their Treevo package is arrange the plant in the mold and hook up the foliage support after potting it in the soil. They can then maintain the plant as if they would a normal plant. When the tree reaches fullness, users remove the elastic mold and continue to water, trim and care for it while keeping its designed shape. "Our ultimate goal is to revolutionize the way cities and countries design their landscape and urban ecosystem by allowing living trees to be used as infrastructures like city lamps, bus stops and more," Said Michael Faber, Head of Design at Treevo (Figure 6.18).

TRUE GARDEN (WWW.TRUEGARDEN.COM)

Grower Arizona (true garden website)

Phoenix, Arizona-based True Garden is the premier vertical aeroponic food farm in the Southwest US. This first-of-its-kind facility, operated by solar power, was designed in partnership with Future Growing LLC (Tower Garden) with a vision to drastically reduce the region's agricultural water consumption while making local, living produce available year-round in the hot desert regions of Phoenix and the Southwest US. True Garden is the first high-tech greenhouse in the US capable of producing most cool season food crops year-round even in the scorching desert, where temperatures reach 120° during the day and 90° at night during peak months. The greenhouse uses very little energy in the winter months by operating in a naturally vented mode. During the hotter summer months, the greenhouse utilizes a combination of smart and efficient technologies to affordably keep the greenhouse at optimal temperatures, both during the day and night. Because of the design of the Tower Garden® system, the crops grow faster than they would in soil, and have to be harvested on a regular basis. Aeroponic towers are lined up in double rows for several hundred feet, creating extra-large plant sites of 120,000–250,000 plants per acre.

As the centerpiece of its state-of-the art greenhouse, True Garden founders Lisa and Troy Albright selected Future Growing's® vertical aeroponic Tower Garden® technology because Future Growing® has a very successful and proven track record—establishing hundreds of urban farms with its vertical aeroponic technology across the country—and the vertical tower farm is ideal for small urban settings like this one. Utilizing Future Growing®'s aeroponic technology which re-circulates valuable water, True Garden uses 95% less water, 90% less space, and no harmful chemical, pesticides, or herbicides like Roundup. The technology benefits urban growers and consumers alike, since the aeroponic plants grown on Tower Gardens® are more nutrient-dense than traditional produce, they grow extremely fast and the Tower Gardens® produce living produce with the roots still intact. You cannot buy produce that is fresher and more hyperlocal than that. True Garden is dedicated to educating the public and empowering homeowners to grow the same way in their own backyard. They sell seedlings and provide consulting services and workshops on aeroponic farming.

REFERENCES

Acoba, E., 6/1/14, *Arizona Daily Star*, https://tucson.com/.../aeroponics.../article_d74927d5-7e98-5771-92fe-7692f641093b.html.

Aero Development Corp website, accessed 9/4/19, www.thinkaero.co.

Aerofarms website, accessed 9/4/19, https://aerofarms.com/farms/.

Aeroponic blog, 7/26/15, Aeroponic Tomatoes, accessed 9/4/19, aeroponicpicture.blogspot.com/2015/07/aeroponic-tomatoes.html.

Aerofarms blog, dell.com, Harnesting IoT to combat food insecurity, waste, and spoilage, https://aerofarms.com/2018/08/13/harnessing-iot-to-combat-food-insecurity-waste-and-spoilage/.

Aerofarms blog, 4/3/18, Plants x Data Science: Phenomics and the Future of Indoor Agriculture, https://aerofarms.com/2018/04/03/plants-x-data-science-phenomics-and-the-future-of-indoor-agriculture/.

Aessensegrows website, accessed 9/7/19, www.aessensegrows.com/zh/corporate/our-story.

AgFunder News, 11/14/14, Aerofarms receives $36 million to convert New Jersey Nightclub into aeroponics farm, www.agrimarketing.com.

Agrihouse website, accessed 9/7/19, www.agrihouse.com.

Amplified Ag website, accessed 9/7/19, www.amplifiedaginc.com.

Andersen, D., 10/30/17, *Durango Herald*, Riverhouse Children's Center begins Aeroponics Project, https://durangoherald.com/articles/191942.

Bazak, E., 1/31/17, *Daily Bruin*, Bruin Plate looks to nurture salad bar greens in aero-ponic towers, https://dailybruin.com/2017/01/31/bruin-plate-looks-to-nurture-salad-bar-greens-in-aeroponic-towers/.

Bibbs, R., 8/19/18, *Herald Bulletin*, Tower garden: Liberty Christian School Garden Club works on Aeroponic gardening, www.heraldbulletin.com/community/tower-garden-liberty-christian-garden-club-works-on-aeroponic-gardening/article_5d428185-139e-58a7-81b2-ff093c2a65c2.html.

Burwood-Taylor, L., 1/17/19, AgFunder, 80 Acres Farms raises 40 million to complete fully automated vertical farm, https://agfundernews.com/80-acres-farms-raises-40m-to-complete-fully-automated-vertical-farm.html.

Chicago Department of Aviation (CDA), 2011, Aeroponic Garden, www.flychicago.com/ohare/ServicesAmenities/amenities/.

Clang, H., 5/10/18, GreenBiz, Why data is an essential nutrient for Aerofarms crops, https://aerofarms.com/2018/08/13/why-data-is-an-essential-nutrient-for-aerofarms-crops/.

Cloudponics website, accessed 9/5/19, www.cloudponics.com.

Devaney, K., 6/23/15, *Miami Herald*, Soil-less fruits and vegetables: The sky is the limit with Aeroponic system, www.miamiherald.com.

Dillet, R., 12/4/18, Agricool raises another 28 million to grow fruists in containers, https://techcrunch.com/2018/12/03/agricool-raises-another-28-million-to-grow-fruits-in-containers/.

Firestone, J., 6/23/15, New Atlas, Podplants: Modular, plug and play vertical gardens for indoor spaces, https://newatlas.com/podplants-indoor-vertical-garden-green-wall/38093/.

Friedman, G., 10/30/14, Bloomberg, www.bloomberg.com/news/articles/2014-10-30/aerofarms-plans-aeroponic-farm-in-newark-to-grow-leafy-greens.

Gibson, P., 6/13/18, aessencegrows blog, www.prnewswire.com/news-releases/aessensegrows-international-indoor-farming-symposium-opens-to-full-house-in-shanghai-300666180.html.

Green Bronx Machine, Oct 2017, Fresh Lessons, Time for KIDS, https://greenbronxmachine.org/press/time-for-kids-fresh-lessons-green-bronx-machine-in-new-york-city-teaches-students-the-skills-to-grow-food-and-profits/.

Green Hygenics Holdings website, accessed 4/17/20, www.greenhygenics.com.

Gro-pod website, accessed 9/7/19, www.gro-pod.co.uk.

Growx website, accessed 9/7/19, www.growx.co.

Hardei, P., 1/19/17, Free Vertical Farm Newsletter, https://urbanverticalfarmingproject.com/2017/01/19/one-time-largest-vertical-farm-shuts-down/.

Helioponix website, accessed 9/7/19, www.heliponix.com/about us.

Higgins, K. T., 6/1/80, Space Station Greens, www.foodengineeringmag.com › articles › 86591-space-station-greens.

Hughes, C. J., 4/7/15, New York Times, www.nytimes.com/2015/04/08/realestate/commercial/in-newark-a-vertical-indoor-farm-helps-anchor-an-areas-revival.html.

igrow, 7/29/19, www.igrow.news/news/indoor-farms-of-america-has-gone-out-of-business

James E Wagner website, accessed 9/7/19, www.jwc.ca.

Kenmore, A., 9/19/17, Watertown Daily News, SUNY Potsdam brings Tower Garden farms to campus, www.watertowndailytimes.com/news05/suny-potsdam-brings-new-tower-farms-on-campus-20170919.

Lettuce Abound website, accessed 9/7/19, www.lettuceabound.com.

Martin, D., 1/31/17, prnewswire, Indoor Farms of America Bringing Fully Off-Grid Containerized Vertical Aeroponic Farms to Market, Plus Media Solutions.

Massey, S., 6/8/18, TEDx Talk, Wabash College.

Nelson, B., 12/12/2007, NBC News, Frontiers: Could vertical farming be the future? Farm capable of feeding 50,000 people could fit 'within a city block', www.nbcnews.com.

Plant factory website, 2018, accessed 9/7/19, www.plantfactorysymposium.com/.

Plenty website, accessed 9/7/19, www.plenty.ag.

Reality Fitness, 5/7/14, Naperville Sun, www.realityfitness.com/single-post/.../Hoping-for-Fruitful-Yield-in-Tower-Garden.

Riviera Creek website, accessed 9/7/19, www.rivieracreek.com.

Rouses Market, 2012, Rouses Markets creates sustainable aeroponic rooftop garden above downtown New Orleans store, www.prnewswire.com/news-releases/rouses-markets-creates-sustainable-aeroponic-rooftop-garden-above-downtown-new-orleans-store-151890995.html.

Smith, C., 11/27/14, TulsaWorld, Hydroponic Scissortail Farms grows fresh, local greens, without the dirt.

Sykes, C., 2/25/15, Business Wire, Indoor Harvest, Corp. Announces Plans to Build Aeroponic Vertical Farm in Houston, Texas.

The Living Greens website, accessed 9/7/19, www.thelivinggreens.com.

Tower Garden by Juice Plus, accessed 9/4/19, www.towergarden.com/blog.authors.html/en/authors/tim-blank.html.

Treevo website, accessed 9/7/19, www.treevo.co.

True Garden website, accessed 9/7/19, www.truegarden.com.

Tuttle, D. R., 12/14/14, Journal Record, Growers say method makes healthy food all year, www.capitalpress.com/nation_world/profit/growers-say-method-makes-healthy-food-all-year/article_f9761d29-1031-553b-aeb8-015dfae5617d.html.

Wright, B., 7/9/15, Daily Press, Aeroponics turning heads and tastes, www.montrosepress.com/.../article_5ff1b58a-2602-11e5-b349-a3df537a26ee.html.

Young, J., 6/1/18, Green Machine, www.pressreader.com/singapore.

7 Practice of Aeroponics

Don't only practice your art, but force your way into its secrets; art deserves that, for it and knowledge can raise man to the Divine.

Ludwig van Beethoven

Aeroponics from a practical standpoint borrows much of its technology, science, and processes from the pioneering work done in the field of hydroponics. Therefore the science of how plants grow, the role and function of roots, the essential nutrients, crop-specific optimization scenarios, the greenhouse, and pest control have all been documented extensively using the hydroponic approach (Jones, 2005a).

AEROPONICS VS. HYDROPONICS

There are significant differences in the application of aeroponics versus hydroponics. The key differences are that in aeroponics the plant roots are suspended in air and not immersed in nutrient solution as is the case for hydroponics. In fact even within the aeroponic approaches, there are differences in presenting the nutrient solution to the roots of the plant.

The initial development of aeroponic technology utilized misting or spray nozzles to coat the roots with an aerosol of nutrient solution. This approach required the use of high-pressure pumps (additional cost) which were necessary to deliver the nutrient solution at a constant rate using spray nozzles as well as the concomitant problem of nozzle clogging. However the work by Tim Blank at Disneyworld and his subsequent Tower Garden approach attempted to eliminate this clogging problem and the need for high-pressure pumps by implementing vertical multifaceted columns and simply pumping the nutrient solution to the top of the column with a tube located inside the column and distributing this solution from the top down. The nutrient solution would cascade inside the column and coat the roots with the proper nutrients. This cascade process could be applied using a time sequence where the pump was on for a fixed period of time and then off. This allowed the roots to absorb more oxygen from the air as the roots were exposed directly to the atmosphere.

The other key difference is that aeroponic systems do not use an inorganic substrate to provide support for the roots of plants, except for a plug of rock wool (silica fibers) that provide a place for seed germination and an anchor point between the stem, leaves, and fruit above the rock wool and the roots below the rock wool and inside the column.

In the hydroponics systems, plants have a need for macro and microchemical elements (Jones, 2005b). Carbon, hydrogen, and oxygen are required by the plants for photosynthesis to occur. This is the process that converts carbon dioxide, water, and oxygen gas into sugars and carbohydrates using the energy of the sunlight. Also nitrogen is necessary for these chemical processes, and it is obtained by root absorption of nitrate and ammonium ions and carried to the stem and leaves via the translocation process. It is believed that there are 15 essential elements required to maintain healthy and vigorous growth by absorption and translocation. These elements are C, H, O, N, P, K, Ca, Mg, B, Cl, Cu, Fe, Mn, Mo, and Zn (Jones, 2005c). Of course all these elements need to be either soluble in the nutrient solution and make contact with the roots or in the case of carbon dioxide come in contact with the leaves.

ROLE OF ROOTS AND THE UPTAKE OF WATER AND NUTRIENTS

The role and function of the roots are noted as well (Jones, 2005d). The roots have two major functions. They physically anchor the plant to the growing medium, and they are the avenue through which water and ions (soluble elements) enter into the plant for redistribution to all parts of the plant. Water uptake is essential for the health of the plant. Also oxygen uptake occurs in the roots. One of the differences between aeroponic and hydroponic growing is that the aeroponic roots are exposed directly to oxygen in the atmosphere, which is around 20% oxygen by volume. So there is plenty of oxygen available to be absorbed by the roots. However, for hydroponics when the roots are submersed in the nutrient solution, the concentration of oxygen in water is very low. This is true because the solubility of oxygen in water is only a few mg/L (ppm). This solubility is temperature dependent and is inversely proportional to temperature. So the higher the temperature, the lower the concentration. At 0°C the solubility of oxygen in water is 14.6 mg/L (ppm), whereas at 35°C it is reduced to 6.9 mg/L (ppm) (Jones, 2005e).

The ion uptake by the roots is selective and depends on these factors (Jones, 2005f).

1. passage through the impermeable liquid layer
2. accumulation against the concentration gradient
3. metabolic energy coupled to such transport
4. mechanism of selectivity
5. vectorial transport

There are five process steps that define this concept of ion absorption. These steps are

1. free space and osmotic volume
2. metabolic transport
3. transport proteins
4. charge balance and stoichiometry
5. transport to the shoot

This is a complex process. There are three things that are believed about ion absorption by roots (Jones, 2005g).

1. the plant is able to take up ions selectively even though the outside concentration and ratio of elements may be quite different than those in the plant.

2. Accumulation of ions by the root occurs across a considerable concentration gradient
3. The absorption of ions by the root requires energy that is generated by cell metabolism

The unique features of an active system of ion absorption involve ion competition, antagonism, and synergism. Root growth is dependent on equipment design for optimum root development, aeration, root surface chemistry, and the effect of temperature on root growth.

ESSENTIAL NUTRIENTS

Looking more closely at the essential elements in the plant itself, there are four specific groups as follows (Jones, 2005h).

GROUP I: C, H, O, N, AND S

These elements are utilized in the form of carbon dioxide, bicarbonate, water, oxygen, nitrate, ammonium, nitrogen, sulfate, and sulfite. The ions are absorbed from the nutrient solution and the gases from the atmosphere. The biochemical functions in the plants are the enzymatic processes and assimilation by oxidation–reduction reactions.

GROUP II: P, B

These elements are utilized in the form of phosphates, boric acid, or borate from the nutrient solution. The biochemical functions in the plant include esterification with alcohol groups in the plants, and the phosphate esters are involved in energy transfer reactions.

GROUP III: K, Mg, Ca, Mn, Cl

These elements are in the form of ions from the nutrient solution. Their biochemical function is nonspecific establishing osmotic potentials and also specific reactions in which the ion brings about optimum conformation of an enzyme protein (enzyme activation). In addition, they provide bridging of reaction partners, balance anions, and control membrane permeability and electropotentials.

GROUP IV: Fe, Cu, Zn, Mo

These elements form ions or chelates in the nutrient solution. Their biochemical function is predominantly in chelated form incorporated in a prosthetic group and enables electron transport by valency charge.

A paper by Mattson and Peters on hydroponic recipes (Mattson et al., 2014) indicates that among the more challenging questions for growers beginning hydroponic/aeroponic production is how to design the crop's fertilizer program. Plants require essential elements in the root zone, including the macronutrients (needed in relatively large quantities) of nitrogen, phosphorus, potassium, sulfur, calcium, and magnesium; and the micronutrients (needed in relatively small quantities) of iron, manganese, zinc, boron, copper, molybdenum, chloride, and nickel. All of these nutrients must be supplied by the nutrient solution, although chloride and nickel aren't included in most recipes, as they're available in sufficient quantities as impurities with the fertilizer.

Fortunately, plants have adapted to growing at a wide range of nutrient concentrations. From a practical standpoint, this means that many different nutrient solution recipes can be used successfully to grow a crop.

They recommend that it is absolutely essential to begin with a laboratory analysis of water. The three main things to note are the alkalinity, the electrical conductivity (EC), and the concentration-specific elements. Alkalinity is a measure of water's ability to neutralize acid.

Water source alkalinity is a much more important number to look at than its pH. The pH is simply a one-time snapshot of how acidic or basic your water is; alkalinity is a measure of its long-lasting pH effect.

EC is a measure of the total dissolved salts, including both essential elements and unwanted contaminants (such as sodium). Therefore, EC is a rough measure of water source purity.

The laboratory water analysis will also tell you which specific essential elements and contaminants are in your water. The concentration of essential elements should be taken into account when preparing your nutrient solution. Often linked with your water alkalinity are considerable levels of Ca, Mg, and S in water. It is important to see whether water contains these important secondary nutrients and at what concentration, then one can supply these nutrients through a fertilizer program if not available in sufficient quantities for your crop's recipe. Sodium and chloride are common contaminants in some waters; ideally these should be less than 50 and 70 ppm, respectively.

Once the water source quality is defined, one can begin to plan a fertilizer strategy specific to the crop of interest. Plant fertilizer concentration needs vary depending on the crop grown, the crop growth stage and environmental conditions. However, for a new grower, a good starting point is to simply develop one recipe that works decently well for a range of crop growth stages and conditions.

An example of nutrient solutions for aeroponic production of leafy greens is shown in Table 7.1.

Table 7.2 shows the concentrations of these nutrient solutions in the units of ppm (parts per million).

It should be noted that most hydroponic recipes call for the use of two or three stock tanks. This is necessary to avoid a nasty precipitate or sludge that will occur when specific nutrients are mixed in the concentrated form. In particular, calcium can combine with phosphates and sulfates to form insoluble precipitates. If one is not mixing these formulas in a concentrated stock solution but rather mixing in a dilute or "ready-to-use" form, one can mix these prescribed amounts into one reservoir containing the final water volume. This requires using a stepwise procedure where

TABLE 7.1

Three Aeroponic Nutrient Solution Formulations Shown for a 100 gal Tank

Nutrient Formulations

Jack's Hydro-FeEd (16-4-17)	Single bag commercial nutrient blend that requires 355 g in 100 gal of water
Jack's Hydroponic (5-12-26) plus calcium nitrate	Tank A made up with 284 g of calcium nitrate (15-0-0) and tank B with 284 g of 5-12-26
Modified Sonneveld solution	Tank A

Tank A
184 g of calcium nitrate trihydrate
14.4 g of ammonium nitrate
167.3 g of potassium nitrate
3.8 g of 10% iron-DPTA

Tank B
51.5 g of potassium dihydrogen phosphate
93.1 g of magnesium sulfate heptahydrate
0.29 g of manganese sulfate monohydrate
0.352 g of boric acid
0.023 g of sodium molybdate dihydrate
0.217 g of zinc sulfate heptahydrate
0.035 g of copper sulfate pentahydrate

Some formulations use two separate solutions (A and B) to avoid precipitation of insoluble salts while some are suitable for mixing as one solution.

TABLE 7.2

Comparison of the Nutrients (ppm) Supplied by Three Different Recipes for Lettuce, Herbs, and Leafy Greens

	Jack's Hydro-FeED (16-4-17)	Jack's Hydroponic (5-12-26)+Calcium Nitrate	Modified Sonneveld's Solution
Nitrogen (N)	150	150	150
Phosphorus (P)	16	39	31
Potassium (K)	132	162	210
Calcium (Ca)	38	139	90
Magnesium (Mg)	14	47	24
Iron (Fe)	2.1	2.3	1.0
Manganese (Mn)	0.47	0.38	0.25
Zinc (Zn)	0.49	0.11	0.13
Boron (B)	0.21	0.38	0.16
Copper (Cu)	0.131	0.113	0.023
Molybdenum (Mo)	0.075	0.075	0.024

each component is added individually and goes into a true solution before one adds the next nutrient. This is where it's very important to pick quality nutrients that are very pure and 100% water soluble.

The made-from-scratch method can be difficult for new or smaller hydroponic growers to manage. One commonly used alternative is a two-bag approach using Jack's Hydroponic (5-12-26) and calcium nitrate (Table 7.1). In this method, Tank B contains 5-12-26 pre-blended formula mixed at a rate to deliver approximately 50–100 ppm N. Tank A contains calcium nitrate at 100–150 ppm N and can also be used to add in some useful crop-specific boosters, such as potassium nitrate or individual micronutrient chelates such as iron-EDTA (ethylenediaminetetraacetic acid), DPTA (diethylenetriaminepentaacetic acid), and EDDHA (ethylenediamine-N,N'-bis(2-hydroxyphenyl)acetic acid).

A relatively new one-bag alternative is Jack's Hydro-FeED (16-4-17). This formula is specifically designed to be used as a one-bag formula to deliver a complete nutrient solution to hydroponic and aeroponically grown crops. It was developed specifically for leafy green and herb growers, but has also seen much success as the main grower formula for tomato, cucumber, and pepper crops. What's unique about this formula is its potentially neutral effect on solution pH, as well as its buffered micronutrient package that also includes the essential blend of iron chelates from EDTA, EDDHA, and DPTA. This formula works well for water types with an alkalinity in the range of 40–200 ppm N.

Understanding nutrient mobility relationships within a plant will greatly enhance how you interpret and use the data generated by a tissue analysis. Young leaves tend to show higher levels of the mobile nutrients (nitrogen and potassium) and lower levels of immobile nutrients (calcium, iron, and manganese). Therefore, samples taken from young leaves can be most useful to diagnose calcium or micronutrient deficiencies. If no particular problem is suspected, testing laboratories typically recommend taking samples from recently matured leaves as a decent representation of what's happening to both new and old growth.

One of the earliest hydroponic nutrient solutions developed was named after the developer, Hoagland. It is the basis for the aeroponic nutrient solutions used today. The concentrations of this solution are given here in mg/L, ppm (Jones, 2005i). It is shown in Table 7.3.

Table 7.4 shows the major element and micronutrient ionic forms and normal concentration ranges found in most nutrient solutions. These are generally what commercial nutrient suppliers provide in their products (Jones, 2005j).

From a practical standpoint these nutrient solutions are made from blends of various salts that contain the anions and cations of interest. Four different vendors are listed in Table 7.5 with the specific formulations in g/100 gal of water as well as the mg/L (ppm) in the final blend for commercial greenhouse vegetable production (Jones, 2005k).

That completes an overview on the practical preparation of nutrient solutions and the relationships between weights and water volumes and actual ppm concentrations in these solutions.

There are several other parameters that need to be fully understood in the operation of an aeroponic growing system. These parameters are pH, temperature (water and air), EC, methods of timing the nutrient delivery system, and lighting if conducting the growing indoors. These topics are covered in the Jones reference (Jones, 2005l).

TABLE 7.3
Hoagland Recipe (ppm)

Nitrogen as nitrate	242
Phosphorus	31
Potassium	232
Calcium	224
Magnesium	49
Sulfur	113
Boron	0.45
Copper	0.02
Manganese	0.50
Molybdenum	0.01
Zinc	0.48

TABLE 7.4
Ranges of Elements in Typical Recipes

	mg/L (ppm)
Major Elements	
Nitrogen, nitrate+ammonium	100–200
Phosphorus mono and diphosphates	15–30
Potassium	100–200
Calcium	200–300
Magnesium	30–80
Sulfur, sulfate	70–150
Micronutrients	
Boron borate	0.30
Copper	0.01–0.10
Iron	2–12
Manganese	0.5–2.0
Molybdenum	0.05
Zinc	0.05–0.50

There is also an extensive chapter on "cropping" that covers growing specific crops, i.e., tomatoes, cucumber, peppers, lettuce, herbs, microgreens, strawberries, green beans, sweet corn, okra, melons, and other types of plants (Jones, 2005m). The author concludes with a brief chapter on pest control that includes insects, integrated pest management, sanitation, and prevention procedures (Jones, 2005n). The Jones book is recommended as an essential resource to anyone who is planning on operating an aeroponics growing system as a reference for the key parameters that are needed to run a technically robust system.

TABLE 7.5

Four Nutrient Blends Shown (g/100 gal and ppm)

	g/100 gal			
	Johnson	Jensen	Larson	Cooper
Chemical				
Potassium nitrate	95	77	67	221
Monopotassium phosphate	54	103	–	99
Potassium magnesium sulfate	–	–	167	–
Potassium sulfate	–	–	130	–
Magnesium sulfate	95	187	–	194
Calcium nitrate	173	189	360	380
Phosphoric acid (75%)	–	–	40 ml	–
Chelated iron (FeDTPA)	9	9.6	12	30
Boric acid	0.5	1.0	2.2	0.6
Copper sulfate	0.01	0.5	–	–
Copper chloride	–	0.05	–	–
Manganese sulfate	0.3	0.9	1.5	2.3
Zinc sulfate	0.04	0.15	0.5	0.17
Molybdic acid	0.005	0.02	0.04	–
Ammonium molybdate	–	–	–	0.14
mg/L ppm				
Major Elements				
Nitrogen	105	106	172	236
Phosphorus	33	62	41	60
Potassium	138	158	300	300
Calcium	85	93	180	185
Magnesium	25	48	48	50
Sulfur	33	64	158	68
Micronutrients				
Boron	0.23	0.46	1.0	0.3
Copper	0.01	0.05	0.3	0.1
Iron	2.3	3.8	3.0	12.0
Manganese	0.26	0.81	1.3	2.0
Molybdenum	0.007	0.03	0.07	0.2
Zinc	0.024	0.09	0.3	0.1

KEY FEATURES OF A COMMERCIAL AEROPONIC SYSTEM

For the purpose of this book a vertical aeroponic system will be described, and it should be noted that the horizontal stacked systems are mainly hydroponic systems although there are some that use misting and therefore are aeroponic systems. This vertical column aeroponic system requires seven key components.

1. Nutrient solution main tank
2. Pump and timer system
3. Column or wall system for staging the plants
4. Transfer lines to deliver the nutrient solution to the plants and return to the main tank
5. Sensors for water and air parameters
6. Water, low total dissolved solids
7. Nutrients

NUTRIENT SOLUTION MAIN TANK

The main tank should be scaled to the capacity of the aeroponic system, i.e., the number of plants to be fed by the tank. A polyethylene tank can be installed above ground or buried below ground. It needs to be accessible for additions of fresh nutrients either manually or using inlet tubing and small accurate delivery pumps. Peristaltic pumps have been used for this purpose.

The nutrients can be obtained from a supplier for the hydroponic operations and the concentrations are chosen for the specific plants to be grown.

The pump and timer systems need to be scaled to the size of the system that it is used with. Typical flowrates are 3–4 gal per minute. Depending on the type of aeroponic system chosen, the pump may need to be a low- or high-pressure pump. A low-pressure system is suggested from both a cost and ease of operation viewpoint. The two major systems are the misted and the non-misted. The misted configuration requires a high-pressure pump and involves spray nozzles. One of the largest drawbacks of these systems is maintaining the flow through the nozzles and the concomitant plugging possibility. The non-misted system uses a low-pressure pump and the nutrient solution is pumped to the top of the column or wall and is distributed as a stream cascading down the inside of the column or wall and coating the roots with nutrients as it returns to the main tank to be recycled.

The pump timer should be chosen to allow for various sequences of ON/OFF that again needs to be determined empirically depending on the plants being grown. Typical sequences are given in minutes. An example would be 15 minutes on followed by 15 minutes off.

There are many possibilities for the different configurations for these systems. The key features include access of nutrients to the roots. The misted systems require that the nozzles be in proximity to the roots so that all the roots get coated by the nutrient mist. For the non-misted system the flow of the cascading water needs to be directed to ensure that the roots get thoroughly coated by the nutrient solution. The other feature is that the area where the roots reside needs to be kept in the dark, so the wall systems need to be enclosed. Typically two-sided walls are constructed back to back with a gap for the nozzles to reside and for the excess water to drain back to the main tank. For the non-misted systems in columns the nutrient solution is directed to flow down the inside of the column, past the roots, and back to the main tank.

Figure 7.1 is a schematic of a trickle-down aeroponic R&D column with 18 plant ports. The submersed pump pumps the nutrient solution to the top of the column, and it cascades down past the rockwool containing the seeds and plants and returns to the reservoir inside the column. Figure 7.2 shows a ten-foot vertical column commercial aeroponic system that has a pump and a reservoir of nutrient solution that is being pumped to the top of the columns and then by gravity flows through the columns and returns to the reservoir for recycling. The plants are located on the sides of the columns so that the roots receive the nutrients so that the plants can receive the necessary light and carbon dioxide from the air. Finally Figure 7.3 shows a picture of an actual growing commercial unit with 8 ten-foot vertical columns showing approximate 600 plants per unit in full growth with lettuce plants.

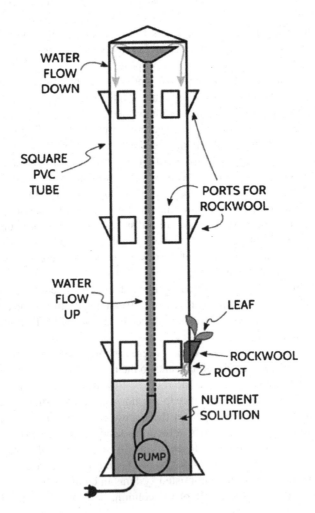

FIGURE 7.1 Schematic of a trickle-down aeroponic column with 18 ports that is used for R&D purposes or for home use. (Reprinted with permission from Aero Development Corp.)

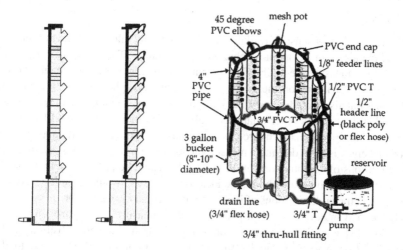

FIGURE 7.2 Vertical column aeroponic commercial unit. (Reprinted with permission from Grow Lode.)

FIGURE 7.3 Photo of commercial aeroponic pod column systems in a greenhouse. (Reprinted with permission from Aero Development Corp.)

TRANSFER LINES

Typically PVC tubing is used to connect the main tank, pumps, and columns or walls and to circulate the nutrient solution throughout the system. This kind of tubing is easy to work with and relatively inexpensive. There have been some concerns raised about the presence of chlorine in PVC, but any residue chlorine that could migrate to the nutrient solution would be in trace concentrations and not be deleterious to the plants.

SENSORS

There are a variety of sensors available for continuous monitoring of the nutrient solution and the air in an aeroponic greenhouse or container. The key sensor used today is designed to measure the total EC of the nutrient solution in microSiemens. This not only provides a window on the concentration of nitrates, phosphates, and potassium (NPK) ions but also includes any other ions present in the solution. Using typical nutrient blends this includes other relatively high concentrations of cations like calcium, magnesium, ammonium, and hydrogen and anions like nitrates, sulfates and chlorides. Unfortunately, it is not a definitive measurement, since for a specific EC reading, there can be millions of combinations of these ions that could be present. As mentioned in the Innovation section of this book, the ideal sensor system would include sensors that measure specific concentrations of individual ions, i.e., a nitrate sensor.

The second major sensor is a pH sensor that measures the hydrogen ion concentration. pH is defined as the negative log of the molar concentration (moles/liter) of hydrogen ions. Typically the nutrient solution should be maintained in the 5.5–6.5 pH range which is slightly acidic. The solubility of the nutrient ions is dependent on the pH. For example high pH conditions will cause calcium phosphate to precipitate, thereby making it unavailable to the roots. The pH is normally maintained by adding small amounts of acidic or basic solutions to adjust the pH to the desired range.

Additional sensors are available and can be used to measure a variety of parameters.

This includes water temperature, air temperature, carbon dioxide concentration in the air, and humidity.

The future of sensors for aeroponic systems will be in the area of specific ion measurements. As mentioned earlier the EC sensor is a rough indication of the ion concentration, but it can be misleading. For example at a specific EC reading the nutrient could contain 0 ppm nitrate and 200 ppm potassium or 200 ppm of nitrate and 0 ppm of potassium. So the need is for a robust and affordable sensor for continuous monitoring of the nutrient composition. Probably the first step in the development of these sensors will be offline analytical tests that would be performed on a daily or weekly basis to ensure that the nitrate, phosphate, and potassium or other major ions are present at the proper level.

Ion-specific sensors have been under development to meet this need for more precise measurement of nutrient solutions. Optical nitrate sensors are one example. Optical nitrate sensors operate on the principle that nitrate ions absorb UV light at wavelengths less than 240 nm. Commercially available, optical nitrate sensors utilize this property to convert spectral absorption properties measured with the sensor to a nitrate concentration by using laboratory calibrations and integrated algorithms to account for interferences from other absorbing ions (such as bromide) and colored organic matter. This allows for real-time nitrate measurements without the need for chemical reagents.

Companies currently manufacturing in situ optical nitrate sensors for industrial and environmental applications include Satlantic (Seabird website), Hach (Hach website), Trios (Trios website) and s::can (s::can website). All manufacturers use a

UV light source and a spectrometer, but instrument designs have important differences in terms of lamp type (deuterium vs. xenon), optical pathlength (1–35 mm), and the process algorithms (two-beam to full UV spectrum) used to calculate nitrate concentrations. Other differences in instruments among manufacturers include the size, geometry, accuracy, detection limit, need for proprietary controllers, maximum sampling rates, and anti-fouling techniques.

A master thesis study focused on ISE (ion selective electrode) development for nutrient solutions (Zhao, 2014). In this thesis a state-of-the-art hydroponic reservoir nutrient controllers were employed for measuring the EC of the solution. This value approximates the reservoir nutrient concentration as a single combined measurement. Studies have shown that plant growth requires over a dozen specific ions for proper nutrition. When a bulk measurement is utilized, nutrient excesses in one ion may mask a deficiency in another. This problem is especially prevalent in reservoirs that utilize nutrient recirculation as part of their reservoir management. To prevent these nutrient deficiencies from arising, and thereby preventing optimal growth, ion-specific electrodes (ISE) were utilized to characterize ionic concentrations. In this new and improved method described in the thesis, the reservoir is polled for concentrations of key ions, rather than obtaining a single bulk measurement. Through real-time monitoring of these data, concentrations of major ions can be determined independently and held at constant rates by a nutrient injection system. Specific objectives of the project involve interfacing ion sensors for nitrate, potassium, pH, and calcium with an embedded controller; these ions are critical for plant growth and significantly influence yield outcomes. These ions were chosen to provide proof of principle within a reasonable time frame and represent the three key macronutrients for plant growth. In addition, concentration data for these ions are used to automatically react and control ionic concentrations.

Reservoir control by a fully automated ion controller may mitigate the effects of heavy ion depletion from growth, as well as keep reservoir ion concentrations held in a steady state. Grower setpoints parameters are software-selectable, which the feedback control system then uses to monitor reservoir concentration for changes. Should the reservoir deviate outside of this grower setpoint by a user-defined percentage, nutrient injection will occur and the program will attempt to rectify any nutrient deficiencies by the addition of a specific ion.

Another study used ion-specific nutrient management systems to monitor nutrient solutions (Barnsey et al., 2012). Plants require a wide range of nutrients to support their growth, development, and reproduction. Each of these specific nutrient ion species has an ionic activity* window within which growth is optimized. Activity, although less commonly used by the greenhouse industry, is related to concentration and is in fact the more important fundamental parameter with respect to plant nutrition. Additionally, it is ion activity, not concentration which ion-selective sensors typically measure. The caveat to this is that all the nutrient ion species need to be within their respective activity windows if plant productivity is to be optimized. Departure from these optimal levels in any of the nutrient ions will have an influence on all the others as well. The uptake and utilization of nutrients depend not only on the absolute quantities but also on the ratios among nutrient species. Deviations above or below these ion activity regimes can lead to the development of toxicity or deficiency symptoms and

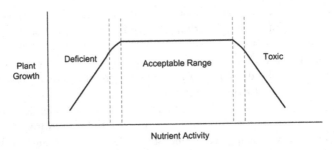

FIGURE 7.4   General principle of plant nutrient acceptable/sufficiency ranges. Dashed vertical lines represent marginal zones between acceptable and deficient or toxic nutrient activity.

ultimately impair productivity. These acceptable nutrient ion activity ranges, which are
often termed sufficiency ranges, can be visualized for a given nutrient ion in Figure 7.4.

Traditionally, the nutritional appropriateness of aeroponic nutrient solutions
used in greenhouse plant production is obtained by monitoring solution pH and EC.
Although this provides some information about the nutrient ions present in the solution, EC is an indiscriminate measure for the total nutrient composition and does
not differentiate among the nutrient species present. Likewise, any non-nutrient ions
within the solution will also contribute to the solution EC, and thus if present, these
ions can result in non-ideal nutrient management choices. Further, different nutrient
ions contribute disproportionately to the measured EC value, which can skew the
interpretations made based on the EC metric (e.g., although potassium and sodium
ions have the same charge, potassium contributes more strongly, for an equivalent
increase in activity, to solution EC. Moreover, in addition to the indirect effect pH
has on EC by inducing precipitation/dissolution reactions, as H^+ and OH^- contribute
differently to solution EC, pH changes can influence EC measurements, further com-
plicating the utilization of EC as a tool for assessing nutrient status.

Plant productivity can be influenced by the activity of any one nutrient ion species.
Given this, pH and EC measurements alone do not provide sufficient information to
allow growers to realize optimal plant production from a solution fertility perspective. Some greenhouse growers do attempt to manage their nutrient solutions based
on individual nutrient species; however, their efforts are often temporally restricted.
Currently growers are limited to relatively infrequent (e.g., 1–3 weeks) offline analysis
in which nutrient solution samples are physically mailed to off-site accredited laboratories. During the shipping, processing, and reporting lag time, the status of the on-farm
nutrient solution will have changed, potentially to a significant degree, limiting the
usefulness of what is often expensive data. Plant tissue/leaf analysis not only serves
as an additional offline analysis technique informing adjustment decisions but also
presents the disadvantage of cost and time lag between sampling and result delivery.

Many growers rely on experience and a keen eye to detect (visually) the symptoms of nutrient deficiency and/or toxicity. Although effective, this method is
reactionary in nature. In most cases, visual symptoms are manifested only after
prolonged periods of growth in a non-optimal nutritional environment. In certain
cases, visual symptoms for different deficiencies/toxicities can be similar, resulting

in misdiagnosis and potentially leading to inappropriate solution modifications that can further exacerbate the problem.

Several companies have off-the-shelf sensors for both ISEs and optical electrodes. Some of these are very expensive (thousands of dollars) but are robust and accurate. Other companies have very inexpensive offline tests based on strip technology and colorimetric tests. The accuracy of these tests is not good, but they may be adequate for trend analysis and for adjusting the nutrient composition. This is a critical area for research and development of robust and affordable sensors for measuring NPK and other major nutrients in real time online.

WATER

The quality of water used for an aeroponic system needs to be understood. The ideal and least expensive source of water is rainwater or cistern water. This requires a water collection tank that can capture rain from the roof of the greenhouse. All aeroponic commercial systems require large quantities of water, so rainwater meets this requirement.

Other natural sources of water can be used, such as a lake or a pond. But any water supply needs to be analyzed to ensure that it meets a minimum standard. Municipal water sources can be used, but they may contain high levels of sodium and chloride ions.

Recommendations from Jones for the maximum concentrations of water used in soil-less growing are shown in Table 7.6 (Jones, 2005o).

Jones also suggests desired levels and upper limits for other related parameters and all the ions, including the micronutrients (Jones, 2005p). Table 7.7 shows these parameters.

The pH of water can vary over a wide range. This is due to the fact that water can absorb CO_2 from the air, which will affect the pH as well as the ionic composition of the water. The uptake of ions by the roots is dependent on the pH, and the pH also affects the solubility of ions. For example, when the pH is too high, phosphates may precipitate out of solution as calcium phosphate.

TABLE 7.6
Maximum Concentrations of Ions in Water Used in Soil-less Growing

Element/Ion	Maximum Concentration (mg/L [ppm])
Chloride (Cl)	50–100
Sodium (Na)	30–50
Carbonate (CO_3)	4.0
Boron (B)	0.7
Iron (Fe)	1.0
Manganese (Mn)	1.0
Zinc (Zn)	1.0

TABLE 7.7
Optimal Parameter Levels for a Hydroponic Growing System

Parameter	Optimum Level	Upper Limit
EC	0.2–0.5 mS/cm	1.5 mS/cm
Total dissolved solids	128–320 ppm	480 ppm
pH	5.4–6.8	7.0
Bicarbonates	40-65 ppm	122 ppm
Hardness ($CaCO_3$ equiv.)	<100 ppm	150 ppm
Sodium	<50 ppm	69 ppm
Chloride	<871 ppm	108 ppm
Nitrogen	<5 ppm	10 ppm
Nitrate	<5 ppm	10 ppm
Ammonium	<5 ppm	10 ppm
Phosphorus	<1 ppm	5 ppm
Phosphate	<1 ppm	5 ppm
Potassium	<10 ppm	20 ppm
Calcium	<60 ppm	120 ppm
Sulfate	<30 ppm	45 ppm
Magnesium	<5 ppm	24 ppm
Manganese	<1 ppm	2 ppm
Iron	<1 ppm	5 ppm
Boron	<0.3 ppm	0.5 ppm
Copper	<0.1 ppm	0.2 ppm
Zinc	<2 ppm	5 ppm
Aluminum	<2 ppm	5 ppm
Fluoride	<1 ppm	1 ppm

REFERENCES

Barnsey, M. et al., 2012, Ion-specific nutrient management in closed systems: The necessity for ion-selective sensors in terrestrial and space-based agriculture and water management systems. *Sensors (Basel)* 12(10):13349–13392.

Hach website, accessed 4/17/20, www.hach.com.

Jones, J. B., 2005a, CRC Press, pp. 1-418. In: *Hydroponics: A Practical Guide for the Soilless Grower*.

Jones, J. B., 2005b, Chapter 5. The essential elements. In: *Hydroponics: A Practical Guide for the Soil-less Grower*.

Jones, J. B., 2005c, *Hydroponics: A Practical Guide for the Soil-less Grower*, p. 13.

Jones, J. B., 2005d, Chapter 4. The plant root. In: *Hydroponics: A Practical Guide for the Soil-less Grower*.

Jones, J. B., 2005e, *Hydroponics: A Practical Guide for the Soil-less Grower*, p. 26.

Jones, J. B., 2005f, *Hydroponics: A Practical Guide for the Soil-less Grower*, p. 21.

Jones, J. B., 2005g, *Hydroponics: A Practical Guide for the Soil-less Grower*, p. 23.

Jones, J. B., 2005h, *Hydroponics: A Practical Guide for the Soil-less Grower*, p. 33.

Jones, J. B., 2005i, *Hydroponics: A Practical Guide for the Soi-lless Grower*, p. 83.
Jones, J. B., 2005j, *Hydroponics: A Practical Guide for the Soil-less Grower*, p. 94.
Jones, J. B., 2005k, *Hydroponics: A Practical Guide for the Soil-less Grower*, p. 99.
Jones, J. B., 2005l, *Hydroponics: A Practical Guide for the Soil-less Grower*, pp. 89–113.
Jones, J. B., 2005m, *Hydroponics: A Practical Guide for the Soil-less Grower*, pp. 167–272.
Jones, J. B., 2005n, *Hydroponics: A Practical Guide for the Soil-less Grower*, pp. 331–337.
Jones, J. B., 2005o, *Hydroponics: A Practical Guide for the Soil-less Grower*, p. 73.
Jones, J. B., 2005p, *Hydroponics: A Practical Guide for the Soil-less Grower*, p. 76.
Mattson, N. et al., 2014, *A Recipe for Hydroponic Success*. www.greenhouse.cornell.edu/crops/factsheets/hydroponic-recipes.pde.
Seabird website, accessed 4/17/20, www.seabird.com.
s::can website, accessed 4/17/20, www.s-can.at.
Trios website, accessed 4/17/20, www.trios.de.
Zhao, M., 2014, Sinking sensors in solution: A novel hydroponic ion control system. Masters Thesis, Georgia Institute of Technology.

8 Aeroponics Current Research

One good test is worth a thousand expert opinions.

Wernher Von Braun

There are many areas of aeroponics that need to be better understood and will require a focused research effort. The key areas are understanding the optimum nutrient compositions for each crop and how to grow these crops as quickly as possible without sacrificing nutritional value of the food.

This optimization process will require understanding the optimum ion composition of the nutrient solution of both the macro and micronutrients. It will also require an understanding of the optimum wavelengths of light necessary for the photosynthesis process for indoor operations. The aeroponic process needs to be fine-tuned. The flow sequence of the nutrient solution that comes in contact with the roots needs to be optimized. A comparison of the misting versus trickle down approach needs to be understood. Also, simple robust affordable analytical methods will need to be developed to characterize the nutrients produced by the plant. This would include the various sugars and starches but also the antioxidants, the vitamins, the polyphenols, the flavonoids, the chlorophyll, and other medicinal components produced by these plants. Because the aeroponic process is essentially a closed system experiment, it can be designed so that all of the critical parameters can be controlled, monitored, and thoroughly studied. This is much more challenging in soil-grown plants due to the complexity that soil adds to the equation. The role of microorganisms and bacteria also needs to be better understood. Because aeroponics can be designed to have excess amounts of nitrate ions, the necessity for nitrification of nitrogen gas is eliminated and the loss of nitrogen via the conversion of nitrates to dinitrogen oxides and other NO_x gases is reduced or eliminated. However, microorganisms may play a unique role in the root uptake on nutrients.

There is research conducted on hydroponic systems but much of that research is not being published. The Plant Factory, which was mentioned in Chapter 1, is not an aeroponics concept but is almost exclusively hydroponic in methodology but uses the horizontal stacked trays as a vertical farming technology. This technology has been developed in Japan, Taiwan, Korea, and China, and many of the findings of the research conducted by these groups are applicable to the aeroponic model. In two books, *The Plant Factory* (Kozai et al., 2016) and the *Smart Plant Factory* (Kozai et al., 2018), Kozai and his colleagues report comprehensive research findings on many relevant topics:

1. Artificial lighting optimization
2. Current status of research

3. Micro- and macro-plant factories with artificial lighting
4. Rooftop production systems
5. Basic physics and physiology of plants—photosynthesis, transpiration, translocation
6. Nutrition and nutrient uptake
7. Tipburn
8. System design, construction, cultivation, and management
9. Education and training
10. Commercial operations
11. Challenges for the next-generation of Plant Factory technology

LETTUCE NUTRIENT STUDY

Some fundamental experiments were conducted at Rutgers University in 2018 to better understand these complexities. These experiments were conducted in pods each containing eight vertical columns that were each ten feet tall. These pods were manufactured by the Aero Development Corp and were located in a greenhouse in New Brunswick, NJ. A team of researchers were involved with these studies (Ayeni et al., 2018).

The study compared two aeroponic pods each containing approximately 600 lettuce plants—150 each of Bibb, Igloo, Scarlet Red, and Pak Choi. These lettuce plants were germinated from seed in the aeroponic column cups in rockwool and grew for 42 days using two different nutrient solutions. In pod A, the electrical conductivity (EC) was maintained at 1,200 mS/cm and in pod B, the EC was maintained at 2,200 mS/cm. The pH was also maintained at 6.5. The pods were located in a greenhouse to utilize natural light. An automated sensor base system (BoxCar) was utilized to monitor the EC, pH, and temperature continuously throughout the study. The results indicate that the lettuce plants grown in the 2,200 EC nutrient solution (pod B) grew 50% more (fresh weight) during the 42 days than from the pod at the lower EC of 1,200 (pod A). The levels of nutrients were monitored both in the leaf tissue samples at 32 and 42 days and the nutrient concentrations were tested at the same interval. In addition to the improved growth rate in pod B, it was noted that the plants almost depleted the levels of nitrates and potassium ions in pod A and the potassium ions in pod B. The phosphates were not absorbed by the roots to a very large extent. As was expected, the uptake of all these nutrients increased because the size of the plants increased. The size of the 42-day plants in order of weight (highest to lowest) were Pak Choi > Igloo > Scarlet Red > Bibb. The plants at the top of the columns were twice as large as the same plants on the bottom of the column. This indicates the impact of light energy on the growth of these plants and possibly that the higher plants absorbed more oxygen than the plants with their roots submerged in the nutrient solution (essentially the hydroponic effect). Each pod has a pump that circulates the nutrient solution to the top of each of the eight columns and passes over the root systems of each plant. The pump was cycled 8 minutes on and 30 minutes off continuously. The pods were located next to each other so that they were exposed to equivalent sunlight. Each pod contained a hydropak from BoxCar Central that had three sensors in the pod—EC, pH, and temperature of water and three additional sensors in air—temperature of air, humidity, and carbon dioxide levels. These sensors were monitored and recorded every minute and the data were available on the internet. The EC and pH were monitored and adjusted every two or three days to ensure that they were held constant through the growth cycle. The seeds were planted in rock wool in the column ports and germinated.

Samples of the leaf tissues were collected after 32 and 42 days from planting and analyzed for the nutrient content. The nutrient solution was also analyzed for nutrient levels at these two intervals.

The pods were maintained by addition of more nutrient solutions, pH adjustments, and additional water on a weekly basis to ensure that the proper EC and pH were achieved. Table 8.1 shows the additions that were made manually to the pod reservoirs to ensure that the proper levels were observed. As expected, the initial changes were minimal when the seedlings sprouted and during their initial growth. As the plants matured, they began to absorb more nutrients and therefore the EC levels were reduced (due to the uptake of nutrients ions) and needed to be replenished. Also, the pH was shifted mainly to a higher pH as more and more hydrogen ions were being absorbed through the roots of these plants. Table 8.1 shows the date of addition and liters or milliliters of the solution added. The water used was city water. The AB dosimeter blends both the A and B together at a ratio of 30 mL of each to 3,785 mL of water. The A and B in mL is the addition of each individual solution concentrates of A or B. pH down and up were added to maintain the pH level at the desired pH. The initial concentrations of NPK in the pods was calculated based on the concentrates of A (5-12-26) and B (15-0-0). The total concentrations in the concentrates are 26,551, 15,217, and 32,975 ppm (NPK, respectively). When the concentrates are diluted from 30 mL to 3,785 L (1 gal) the final concentrations are 210, 120, and 261 ppm (NPK, respectively). This diluted solution was then added to the 151 L pod reservoirs to achieve an EC of 1,200 in pod A and 2,200 in pod B. In Table 8.1, the designation AB is the final diluted nutrient solution and the designation A and/or B is the concentrate described above, i.e., A is 6,340, 15,217, and 32,975 ppm (NPK, respectively) and B is 20,211 ppm N only.

The nutrient water samples that were taken on the 32nd (5/17/18) and 42nd (5/24/18) day of the study were analyzed for all the nutrients and these data are shown in Table 8.2. The key elements NPK show dramatic differences between pod A and B for both nitrate and potassium. It appears that pod A is almost depleted in nitrogen and potassium during the ten days between the samples. This includes addition of more nutrients between those days. The phosphorus appears to be stable and even shows an increase from day 32 to day 42 due to the addition of fresh nutrients. Therefore, apparently the plants are not absorbing phosphates to the same extent or they are unable to absorb them effectively. One would expect that all these nutrients would be reduced during the growing cycle. The potassium ions appear to be absorbed to a large extent as well. It should be noted that these levels of nutrients were being maintained using EC as the control parameter. This points out how inadequate this measurement is for actually understanding the actual levels of ions. In this example although the EC was maintained at 1,200 and 2,200 mS/cm, in reality

TABLE 8.1
Water, Nutrient Solution, and pH Adjustment Additions

	Pod	Water (L)	AB (mL)	A (mL)	B (mL)	pH down (mL)	pH up (mL)
4/16/18	A	13.2					
4/17/18	A					600	
	B			20	20		
4/18/18	B						40
4/20/18	A					150	
	B		6.3	20	20		
4/23/18	A	31.5					
	B	31.5					
4/24/18	A	10.5				250	
	B		10.5				
4/27/18	A		10.5			250	
	B		21				
4/30/18	A	21				1,100	
	B	10.5					30
5/7/18	A	10.5	21			2,000	
	B	10.5	21			1,850	
5/8/18	A		10.5			1,500	
	B		10.5			1,600	
5/9/18	A		10.5			850	
	B		10.5			1,000	
5/14/18	A	10.5	31.5				
	B		46.2	110	110	2,250	
5/15/18	A	25.2				1,500	
5/17/18	A		10.5	10	10	750	
	B		12.6	130	130	500	
5/18/18	A		10.5			500	
	B		10.5	150	150		
5/19/18	A		21				
	B		16.8	20	20		
5/21/18	A		12.6				
	B		25.2	70	70		
5/22/18	A	10.5	10.5			500	
	B		18.9	100	100	500	
5/23/18	A		18.9	30	30	500	
	B	10.5	16.8	30	30	500	
5/24/18	A		37.8	50	50		
	B		44.1	190	190	1,250	

TABLE 8.2
Nutrient Concentration in ppm in the Nutrient Solution at 32 and 42 days

	NO$_3$–N	P	K	Ca	Mg	Na	S	Cl	Fe	Mn	B	Cu	Zn	Mo	Al
							32 days								
POD A	7.67	82.06	4.05	95.79	39.64	99.52	60.41	156.99	1.12	0.03	0.20	0.14	0.08	0.04	0.10
POD B	59.80	80.75	60.26	160.65	92.40	84.29	138.70	125.80	3.00	0.09	0.64	0.23	0.15	0.09	0.09
	52.13	–1.31	56.21	64.86	52.76	–15.23	78.29	–31.19	1.88	0.06	0.44	0.09	0.07	0.05	–0.01
							42 days								
POD A	0.12	98.98	0.19	106.66	51.97	77.72	88.93	108.59	1.54	0.00	0.20	0.21	0.08	0.36	0.02
POD B	42.69	100.83	3.00	222.18	150.30	94.75	250.05	125.35	4.15	0.04	1.03	0.45	0.17	0.28	0.01
							Differences								
POD A 42–32	–7.55	16.92	–3.86	10.87	12.33	–21.80	28.52	–48.40	0.42	–0.03	0.00	0.07	0.00	0.32	–0.08
POD B 42–32	–17.11	20.08	–57.26	61.53	57.90	10.46	111.35	–0.50	1.15	–0.05	0.39	0.22	0.02	0.19	–0.08

the phosphate levels were increasing as more nutrient mix was added to the pods while the nitrates and potassium were depleted substantially and even eliminated. Therefore, the operator thinks that the nitrates are adequate for the plants, but in reality, it is mostly taken up by the plants and only phosphates are present in the nutrient solution.

The other macronutrients (Ca, Mg, Na, S, and Cl) are absorbed at different rates. Sodium and chloride are depleted between day 32 and 42 due to uptake by the plants in pod A. However, pod B shows a slight increase for sodium and no change for chloride. Calcium, magnesium, and sulfur are increased between day 32 and 42 for both pods A and B.

The micronutrients (Fe, Mn, B, Cu, Zn, Mo, and Al) show minor changes and they are present at much lower concentrations, therefore, the magnitude of any change is difficult to measure accurately. Iron increases in both pod A and B indicating that the plants are not absorbing these ions readily.

LEAF TISSUE SAMPLES

The leaf tissue samples were taken on the 32nd and 42nd day of the study and analyzed for nutrient concentration. Each plant was analyzed in duplicate and the results are compiled in Tables 8.3–8.6.

The Bibb lettuce tissue samples show large differences between the A and B plants. Pod B shows higher levels (wt%) of N and K and slightly higher levels of P after 32 days than pod A. However, it decreases in N and K between 32 and 42 days and shows minor differences for P. The comparison of 32 to 42 days for pod A shows increased sodium and manganese levels. In pod B, only the iron levels show an

TABLE 8.3
Bibb Leaf Tissue Sample Analysis

	Days	32	32	42	42	42-32	42–32	32	42
	POD	A	B	A	B	A	B	B–A	B–A
	Conc								
N	%	3.07	4.75	3.03	4.82	--0.04	0.07	1.68	1.79
P	%	0.52	0.61	0.71	0.82	0.19	0.21	0.09	0.11
K	%	5.3	8.67	3.57	6.77	−1.73	−1.9	3.37	3.2
S	%	0.22	0.29	0.27	0.33	0.05	0.04	0.07	0.06
Mg	%	0.38	0.47	0.49	0.65	0.11	0.18	0.09	0.16
Ca	%	1.08	1.18	1.23	1.25	0.15	0.07	0.1	0.02
Na	%	0.51	0.31	1.64	0.67	1.13	0.36	−0.2	−1
B	ppm	20	22	32	32	12	10	2	0
Zn	ppm	20	20	26	29	6	9	0	3
Mn	ppm	56	127	107	121	51	−6	71	14
Fe	ppm	67	74	85	89	18	15	7	4
Cu	ppm	5	5	9	8	4	3	0	−1
Al	ppm	63	34	48	33	−15	−1	−29	−15

TABLE 8.4

Igloo Leaf Tissue Sample Analysis

	Days	32	32	42	42	42–32	42–32	32	42
	POD	A	B	A	B	A	B	B–A	B–A
	Conc								
N	%	2.9	5	2.4	5.2	−0.5	0.2	2.1	2.8
P	%	0.4	0.6	0.6	0.8	0.2	0.2	0.2	0.2
K	%	4.5	85	4.3	9.1	−0.2	−75.9	80.5	4.8
S	%	0.2	0.3	0.2	0.3	0	0	0.1	0.1
Mg	%	0.3	0.4	0.5	0.6	0.2	0.2	0.1	0.1
Ca	%	1	1.1	1.11	1.3	0.11	0.2	0.1	0.19
Na	%	0.3	0.2	1.2	0.4	0.9	0.2	−0.1	−0.8
B	ppm	21	24	31	34	10	10	3	3
Zn	ppm	21	22	24	37	3	15	1	13
Mn	ppm	56	141	63	156	7	15	85	93
Fe	ppm	46	64	60	85	14	21	18	25
Cu	ppm	10	7	6	8	−4	1	−3	2
Al	ppm	21	9	19	35	−2	26	−12	16

TABLE 8.5

Scarlet Red Leaf Tissue Sample Analysis

	Days	32	32	42	42	42–32	42–32	32	42
	POD	A	B	A	B	A	B	B–A	B–A
	Conc								
N	%	3.1	4.54	3.13	4.59	0.03	0.05	1.44	1.46
P	%	0.58	0.65	0.86	0.91	0.28	0.26	0.07	0.05
K	%	5.77	8.33	4.33	7.74	−1.44	−0.59	2.56	3.41
S	%	0.19	0.25	0.24	0.27	0.05	0.02	0.06	0.03
Mg	%	0.31	0.37	0.46	0.46	0.15	0.09	0.06	0
Ca	%	0.82	0.9	1	0.84	0.18	−0.06	0.08	0.16
Na	%	0.36	0.22	1.15	0.37	0.79	0.15	0.14	0.78
B	ppm	20	21	29	30	9	9	1	1
Zn	ppm	19	20	29	32	10	12	1	3
Mn	ppm	54	145	75	115	21	−30	91	40
Fe	ppm	64	69	80	90	16	21	5	10
Cu	ppm	6	6	8	10	2	4	0	2
Al	ppm	58	26	42	32	−16	6	−32	−10

increase. Comparing B to A for each time interval, at day 32 the potassium and manganese levels show an increase. At day 42, nitrogen and manganese levels increase, whereas the aluminum level decreases.

For the Igloo plants, the levels of N and K are much higher in the plants in pod B than pod A and the level of P shows an increase as well. However, N and K levels

TABLE 8.6

Pak Choi Leaf Tissue Sample Analysis

	Days	32	32	42	42	42–32	42–32	32	42
	POD	A	B	A	B	A	B	B–A	B–A
	Conc								
N	%	5.43	7.38	5.15	6.96	0.28	0.42	1.95	1.81
P	%	0.7	0.68	0.91	0.86	0.21	0.18	0.02	0.05
K	%	5.01	5.59	5.38	6.59	0.37	1	0.58	1.21
S	%	1.16	1.05	1.22	1.21	0.06	0.16	−0.11	−0.01
Mg	%	0.66	0.89	0.85	1.24	0.19	0.35	0.23	0.39
Ca	%	3.4	3.62	4.26	5.22	0.86	1.6	0.22	0.96
Na	%	0.55	0.26	1.53	0.56	0.98	0.3	−0.29	−0.97
B	ppm	97	71	111	81	14	10	−26	−30
Zn	ppm	64	33	67	33	3	0	−31	−34
Mn	ppm	106	186	122	188	16	2	80	66
Fe	ppm	54	67	60	81	6	14	13	21
Cu	ppm	8	8	9	8	1	0	0	−1
Al	ppm	23	17	41	42	18	25	−6	1

decrease in pod A from 32 to 42 days, whereas they show an increase in pod B from 32 to 42 days. The higher EC nutrient solution causes the tissue to take up more ions than the lower EC nutrient solution. The comparison of 32 to 42 days for pod A shows increased boron and iron levels. In pod B, only the iron and aluminum levels show an increase. Comparing B to A for each time interval, at day 32 the iron and manganese levels show an increase and the aluminum level shows a decrease. For day 42, zinc, manganese, iron, and aluminum levels increase.

For the Scarlet Red, the NPK levels increase between pod A and B from 32 to 42 days. But minor differences are observed comparing changes from 32 to 42 days within each pod.

The comparison of 32 to 42 days for pod A shows increase in manganese and iron levels and decrease in aluminum levels. In pod B, only the iron levels show an increase, whereas manganese levels show a decrease. Comparing B to A for each time interval, at day 32 and day 42 the manganese levels show an increase and aluminum level decreases.

The Pak Choi data show increase in all of the macronutrients from day 32 to 42 for both pods A and B. Therefore, the Pak Choi continues to take up those nutrients during the growing cycle. The comparison of 32 to 42 days for pod A shows increases in boron, manganese, and aluminum levels. In pod B, boron, manganese, and aluminum levels show an increase. Comparing B to A for each time interval, at day 32 and day 42 the manganese and iron levels show an increase and both boron and zinc show a decrease.

The vertical columns are shown in Figures 8.1 and 8.2 taken at the 32nd day and the 42nd day.

FIGURE 8.1   Thirty-two day photo of pod.

FIGURE 8.2   Forty-two day photo of a pod.

A follow-up study was conducted using the two pods which involved growing Scarlet Red lettuce in pod A and Igloo lettuce in pod B. The objective of the study was to monitor the uptake of the macro and micronutrients during a 45-day study by analyzing a nutrient solution water sample at regular intervals through the growth cycle. The EC for the pods was initially targeted for 1,200 mS/cm for the initial germination and seedling stage, and then, the EC was increased to 2,500 mS/cm on day 11 and not adjusted for the duration of the study. However, pod A did have one adjustment on day 25 which raised the levels of the nutrients slightly after that day.

The data obtained from this study are shown in the following plots of groups of macro and micronutrients.

The NPK levels in the nutrient solution for pod A and B are shown in the figures below.

Figure 8.3 shows the levels of NPK in the Scarlet Red in pod A. The initial levels of the 1,200 mS/cm EC nutrient solution are below 50 ppm for all three ions. When the EC is increased to 2,500 mS/cm, the levels of N, P, and K are increased to 175, 50, and 190, respectively. Then the P slowly decreases to less than 5 ppm as the plant takes up this ion. The N and K appear to remain fairly constant from day 10 to day 32 which indicates that the young plants are not taking up these ions. Then on day 46, the levels of these ions drop by over 100 ppm as the plant absorbs these ions. Additional nutrients were added to this pod at day 25 which affected those levels after that day.

Figure 8.4 shows the levels of NPK in the Igloo in pod B.

This plot shows some similarities and differences between Igloo and Scarlet Red lettuce nutrient uptake. The P levels appear to be relatively similar. However, the P levels are consumed much quicker, and by day 25, the ppm levels of P are almost completely gone. The K levels are also consumed much faster. By day 32, they are almost completely consumed. The N levels follow the same trend as the Scarlet Red

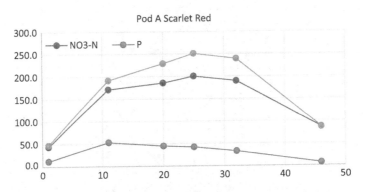

FIGURE 8.3 Levels of NPK in ppm during the growth cycle of the Scarlet Red lettuce.

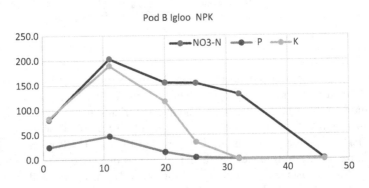

FIGURE 8.4 Levels of NPK in ppm during the growth cycle of the Igloo lettuce.

but even they are reduced more quickly than the Scarlet Red. It would be interesting to measure the levels of chlorophyll in these leaves during the growth cycle to see whether there is a correlation between uptake of N and chlorophyll concentration.

For the other macronutrient levels, Ca, Mg, S, Na, and Cl, are plotted for Scarlet Red in Figure 8.5.

The initial concentrations are all less than 100 ppm for the 1,200 mS/cm solution and then increase when the EC is increased to 2500 mS/cm. The chloride ions are present at the highest levels and remain fairly constant until after day 32 when they are reduced from 200 to 135 ppm. The other ions remain fairly constant and increase slightly which indicates that the Scarlet Red lettuce is not absorbing these nutrients.

Figure 8.6 shows the data for the same micronutrients for pod B—the Igloo lettuce.

For the Igloo lettuce, similar trends are observed. The chloride ppm levels seem to increase during the 10–30-day period and then are absorbed in the final stages of growth. The S levels seem to increase after day 20, more than doubling in

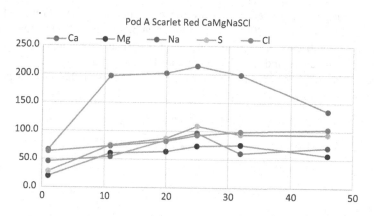

FIGURE 8.5 Levels of Ca, Mg, S, Na, and Cl in ppm during the growth cycle of the Scarlet Red lettuce.

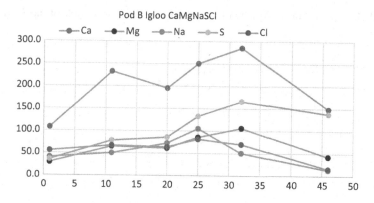

FIGURE 8.6 Levels of CaMgSNaCl in ppm during the growth cycle of the Igloo lettuce.

concentration. Does the root hair chemistry produce more S? The other macronutrients appear fairly constant until the last ten days where they decrease by a factor of 2 indicating absorption by the plants.

Looking at the micronutrients, it should be noted that these ions are present in very low concentrations and so the data may be skewed due to the accuracy of the data.

Figure 8.7 shows the profile for Fe, Mn, and B.

The Fe levels show a major increase from the initial nutrient solution at 1,200 mS/cm to the 2,500 mS/cm solution on day 11. The Fe is fairly constant throughout the growth cycle until the last ten days when the level drops from 2 to 1 ppm. The Mn and B levels appear to be fairly constant and are not being absorbed by the plant.

Figure 8.8 shows the same data for Igloo lettuce.

For Igloo the Fe data are very similar except the final data point that indicates that more Fe is absorbed by the Igloo plants than Scarlet Red. The Mn and B levels are similar in that neither are absorbed to any large extent.

Finally, the CuZnMoAl micronutrients are plotted in Figure 8.9 for Scarlet Red lettuce.

The aluminum ion concentrations are the most unusual in this group. The initial concentration in the 1,200 mS/cm solution appears much higher than any of the other micronutrients. It remains constant at day 11 and then appears to be consumed between day 10 and 25. The pH on day 11, 20, and 25 was 6.2, 6.7, and 6.8. Therefore, pH does not seem to be the cause. Are Al ions chelating with proteins

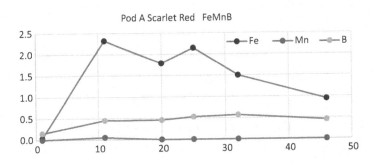

FIGURE 8.7 Levels of Fe, Mn, and B in ppm during the growth cycle of the Scarlet Red lettuce.

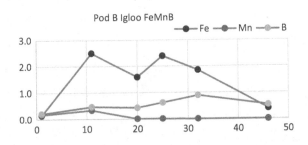

FIGURE 8.8 Levels of Fe, Mn, and B in ppm during the growth cycle of the Igloo lettuce.

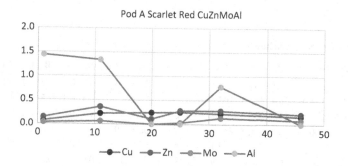

FIGURE 8.9 Levels of Cu, Zn, Mo, and Al in ppm during the growth cycle of the Scarlet Red lettuce.

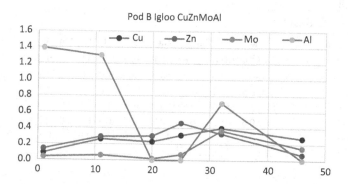

FIGURE 8.10 Levels of Cu, Zn, Mo, and Al in ppm during the growth cycle of the Igloo lettuce.

at the root hairs and then being released before day 30? The Al chemistry is very interesting. The other micronutrients in this group remain fairly constant and do not seem to be consumed by the Scarlet Red lettuce.

Figure 8.10 shows the plot of these micronutrients in the Igloo lettuce pod.

The Igloo pod shows the same strange profile for Al as the Scarlet Red pod. The pH for this pod at day 11, 20, and 25 was 6.0, 7.1, and 8.0, respectively. This would suggest that there is a possibility for the Al ions to form the hydroxide and precipitate. However, the pH on day 32 was 8.2 and Al is present at 0.7 ppm. The other ions seem to remain fairly constant and even increase from day 20 to day 32 before decreasing at the end of the growth cycle.

Tables 8.7 and 8.8 show the actual raw data for these studies.

This study shows the value of the aeroponic experiment to track the ions during the growth cycle of a plant to see which nutrients does this plant absorb and if it does not absorb a nutrient, is that nutrient necessary for the plant. Using these data, one could compare the nutrient solution composition to the synthetic output of the plant, i.e., vitamins, phytochemicals, antioxidants, etc. to optimize the nutrient solution to produce the most nutritious plant.

TABLE 8.7

Pod A Scarlet Red Dataset

Day	pH	EC	NO$_3$–N	P	K	Ca	Mg	Na	S	Cl	Fe	Mn	B	Cu	Zn	Mo	Al
1	7.1	0.8	43.7	11.8	47.5	66.4	19.6	46.0	27.8	63.6	0.0	0.0	0.2	0.1	0.1	0.0	1.4
11	6.2	2.1	171.4	53.4	191.5	197.0	60.8	54.5	74.8	73.2	2.3	0.1	0.5	0.2	0.4	0.1	1.3
20	6.7	2.3	185.5	44.9	228.6	202.2	63.9	84.0	87.2	82.1	1.8	0.0	0.5	0.2	0.1	0.0	0.0
25	6.8	2.4	200.4	42.2	250.8	214.6	73.7	96.6	108.8	92.6	2.2	0.0	0.5	0.3	0.3	0.0	0.0
32	7.0	2.3	189.7	32.8	239.1	199.1	75.7	61.1	94.0	98.9	1.5	0.0	0.6	0.2	0.3	0.1	0.8
46	7.8	1.4	86.5	6.8	86.8	134.9	57.1	70.8	93.7	103.0	0.9	0.0	0.5	0.2	0.2	0.1	0.0

TABLE 8.8

Pod B Igloo Dataset

Day	pH	EC	NO$_3$–N	P	K	Ca	Mg	Na	S	Cl	Fe	Mn	B	Cu	Zn	Mo	Al
1	6.7	1.2	79.9	25.0	82.9	107.6	29.6	41.1	35.9	55.6	0.1	0.2	0.2	0.1	0.1	0.0	1.4
11	6.0	2.3	203.0	47.0	189.0	232.0	65.0	50.0	78.0	67.0	2.5	0.3	0.5	0.3	0.3	0.1	1.3
20	7.1	1.9	155.1	14.8	116.5	195.2	61.5	72.5	85.6	64.0	1.6	0.0	0.4	0.2	0.3	0.0	0.0
25	8.0	2.0	153.9	4.1	34.7	250.3	85.8	105.2	132.8	81.5	2.4	0.0	0.6	0.3	0.5	0.1	0.0
32	8.2	2.1	131.2	1.4	0.6	285.3	106.4	50.9	166.0	70.3	1.9	0.0	0.9	0.4	0.3	0.4	0.7
46	8.4	0.8	1.4	1.3	0.0	148.5	42.9	13.3	138.0	15.8	0.4	0.0	0.5	0.3	0.1	0.2	0.0

KALE PHOTOPERIOD STUDY

An aeroponic kale photoperiod study was conducted at Charleston Southern University (Apr–Aug 2019) to determine the optimum photoperiod sequence for optimizing the fresh weight of kale using four 32″ 18 port Aero Development Corp towers. This study was designed to optimize the cycle time for growing kale dwarf blue curled vates. The study time was set for 28 days from seed planting to harvest. All the variables were held constant except for the photoperiod sequence.

EXPERIMENTAL

Four different sequences were used. These are shown in Table 8.9.

The four aeroponic towers had four different types of lighting configurations. Each tower was given a name. The names were Gertrude, Hannah, Sophia, and Tatiana. The lighting configurations are given in Table 8.10.

The lights were attached to a PVC (polyvinylchloride) tubing and the distance between the light strips and the tower port was fixed. Gertrude and Tatiana had lights that were 12″ from the ports and Hannah and Sophia had lights that were 6″ from the ports. The ports were numbered from the bottom of the tower to the top. There are three levels of ports (bottom/middle/top). Ports 1–6 are on the bottom, 7–12 in the middle, and 13–18 on the top. Port 1 was designated as the water fill port and therefore

TABLE 8.9

Aeroponic Photoperiod Sequences

Cycle	Hours ON/OFF	Hours ON for One Day	Date Range
1	10/14	10	4/24–5/22
2	7/1	21	5/22–6/19
3	3.5/0.5	21	6/19–7/17
4	4/4	12	7/17–8/14

TABLE 8.10

Lighting Configurations for the Four Towers

Tower	Type of Light	Mfgr	Watts	Wavelength	Strip Length Inches
Gertrude	T5 LED	Monios	10	Full spectrum	24
Hannah	T5 LED	Monios	10	Full spectrum	24
Sophia	LED	Litever	45	410–470/600–800 nm	16
Tatiana	LED	Litever	45	410–470/600–800 nm	16

no kale was planted in that port. The photosynthetic active radiation (PAR) was measured using an Apogee light meter. The readings were given in micromoles m^2/s^{-1}.

These readings were taken on 8/14 and the end of the study and are compiled in Table 8.11.

These data are represented graphically in Figure 8.11.

The nutrient solution used for ALL the experiments was J R Peters A and B solid nutrient. Nutrient A contained an NPK ratio of 5-12-26 and nutrient B contained 15-0-0. Each tower nutrient solution was prepared by dissolving 4 g of A and 4 g of B in 8 L of tap water.

These initial fresh solutions were analyzed by J R Peter Lab to determine the exact concentrations of each ion present. Table 8.12 shows the elemental composition of the tap water and the initial nutrient solutions used in ALL the experiments conducted.

The seeds (Johnny Seeds) are planted in 1″ square rockwool cubes and the cubes are placed in the ports. The pump used in each tower is a submersible EcoPlus with a flow rate of 8.2 L/min. The pumps were controlled by a timer that used an on/off sequence of 15 minutes ON and 15 minutes OFF continuously.

The pumps pump the nutrient solution to the top of the tower, and it cascades by gravity down the inside walls of the tower soaking the rockwool and providing nutrient solution to the seeds initially and the roots as the kale plants sprout and grow. These solutions were adjusted to maintain a pH in each week in the range of 5.5–6.5. In each week the volume of the tower was topped off to approximately 8 L using tap water to replace any water lost by evaporation or transpiration. The pH and EC measurements were then taken after the top off was completed. At the end of each 28-day cycle, the nutrient solution was sampled and sent to J R Peters Lab for elemental analysis. A fresh nutrient solution was then prepared at the beginning of

TABLE 8.11

Photosynthetic Active Radiation (micromoles m²/s⁻¹) at Each of the 18 Tower Ports

Port Number	Hannah	Gertrude	Tatiana	Sophia
1	39	20	23	15
2	84	43	42	7
3	88	87	60	7
4	67	57	25	14
5	80	104	40	12
6	18	22	42	12
7	76	194	76	160
8	86	47	73	70
9	90	135	101	146
10	54	100	87	63
11	102	142	68	155
12	52	141	81	88
13	61	20	37	111
14	69	51	52	82
15	74	147	69	146
16	78	69	42	139
17	95	192	47	112
18	58	30	70	107

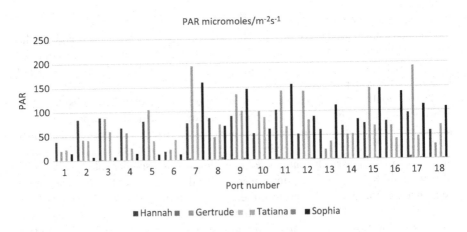

FIGURE 8.11 PAR readings for the 18 ports in the four towers.

each cycle. The EC was measured each week as an additional check of the nutrient concentration. EC is a measure of the conductivity of all the ions present, therefore, it provides a bulk measurement but does not indicate what specific ions are present in the solution. The Peters elemental analysis provides that information on day 1 and day 28 of the experiment.

TABLE 8.12
Concentrations of Ions in ppm (parts-per-million or mg/L)

Ion	Tap Water (ppm)	Fresh Nutrients (ppm)
K	0.99	88.23
Ca	17.11	90.37
Mg	2.14	26.76
Na	7.57	9.96
P	1.29	34.08
S	6.86	40.27
Cl	14.88	15.07
NH_4-N	0.67	5.48
NO_3-N	0.25	83.92
Fe	0.03	0.76
Mn	0.01	1.16
B	0	0.15
Cu	0.04	0.26
Zn	0	0.26
Mo	0	0.18
Al	0	0.41

KALE RESULTS

The average fresh weights of the kale are shown in Figure 8.12. It should be noted that the germination rates were not 100% in each cycle, therefore, the average weight was calculated based on the total number of weighed mature kale plants.

The PAR for each tower is shown in Figures 8.13–8.16.

The correlation of PAR and fresh weight was compiled for the 7:1 and the 4:4 photoperiod. These data are shown for all four towers in Figures 8.17–8.20 (4:4) and Figures 8.21–8.24 (7:1). The linear square fit equation and the correlation coefficient, R^2, are shown in each plot.

The nutrient solution data are shown for the four different photoperiod cycles in Figures 8.25–8.28.

The data clearly show that some photoperiods produce better yields. Especially, the 7 hour on 1 off sequence shows the yields for Hannah and Gertrude. Hannah was particularly dramatic.

Comparing the uptake of NPK for this photoperiod for Hannah, it should be noted that the N and K concentrations declined to near zero which means that as the plants grew, they required much more N and K. Even the P concentrations were much lower than the other experiments. This was also true for the Mg, Ca, and S concentrations which decreased more than in the other experiments as well. This same trend was observed for NPK and MgCaS in the 3.5-hour sequence but not as dramatic as in the 7.1-hour sequence.

Comparing the 7 hour on 1 hour off to the 3.5 hour on 0.5 hour off sequence, it should be noted that they both provide 21 hours of light during 24 hours. However,

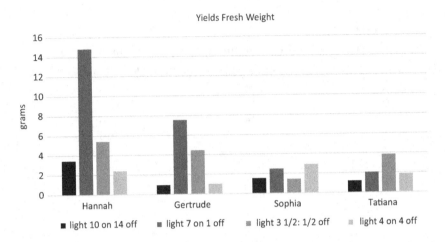

FIGURE 8.12 Average fresh weight of the kale plants for ALL four cycles and ALL four towers.

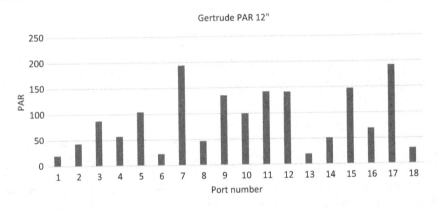

FIGURE 8.13 PAR for Gertrude.

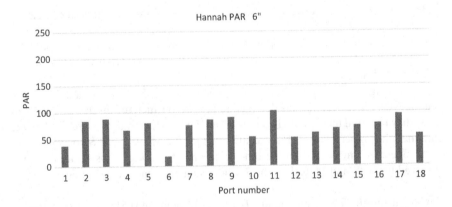

FIGURE 8.14 PAR for Hannah.

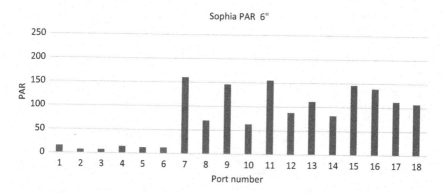

FIGURE 8.15   PAR for Sophia.

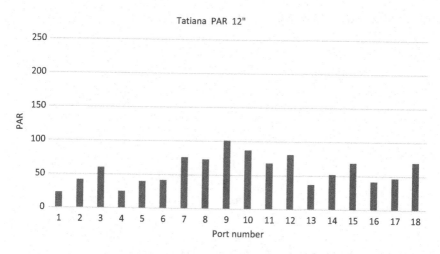

FIGURE 8.16   PAR for Tatiana.

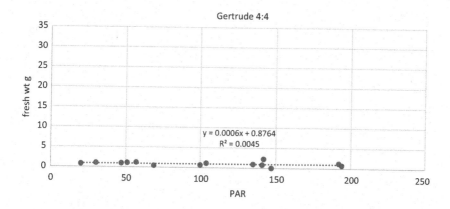

FIGURE 8.17   PAR vs. fresh weight for Gertrude 4:4 photoperiod.

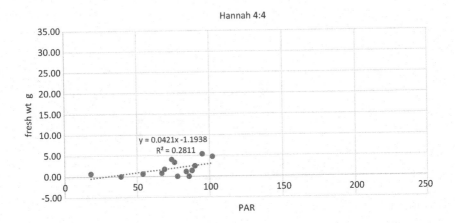

FIGURE 8.18   PAR vs. fresh weight for Hannah 4:4 photoperiod.

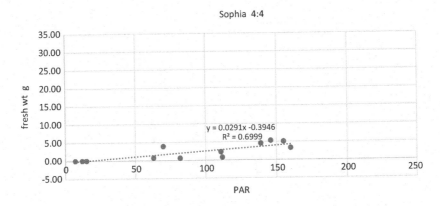

FIGURE 8.19   PAR vs. fresh weight for Sophia 4:4 photoperiod.

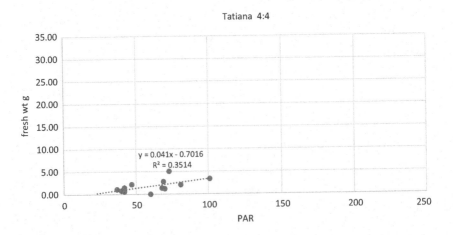

FIGURE 8.20   PAR vs. fresh weight for Tatiana 4:4 photoperiod.

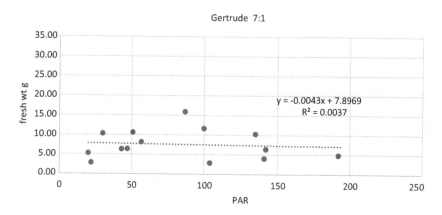

FIGURE 8.21 PAR vs. fresh weight for Gertrude 7:1 photoperiod.

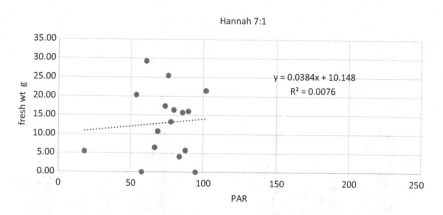

FIGURE 8.22 PAR vs. fresh weight for Hannah 7:1 photoperiod.

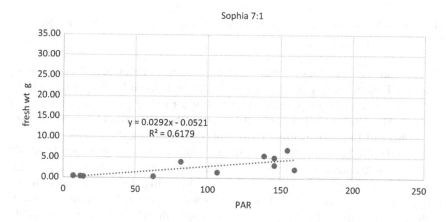

FIGURE 8.23 PAR vs. fresh weight for Sophia 7:1 photoperiod.

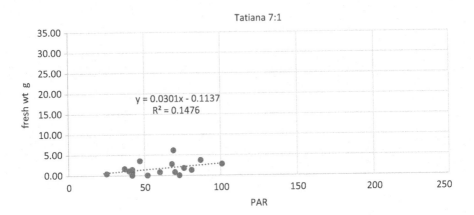

FIGURE 8.24   PAR vs. fresh weight for Tatiana 7:1 photoperiod.

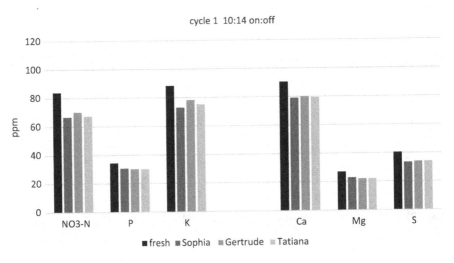

FIGURE 8.25   Nutrient solution ppm concentrations (fresh and at 28 days) for NPK and MgCaS for cycle 1.

the 3.5-hour sequence shows lower yields for Gertrude, Hannah, and even Sophia. Whereas Tatiana better yields for 3.5 vs. 7. The conclusion is that possibly during the 7-hour photosynthetic process, there is a higher rate of growth as more time is allowed. Also, the wavelengths of light missing in the LED strips (470–600 nm) used for Sophia and Tatiana may impact different types of chemical reactions.

It should also be noted that the PAR meter measures a total radiation from 400 to 800 nm so that measure is a composite of all wavelengths. On that note, the LED strips were attached to the PVC frame at the beginning of cycle 1 and were not changed throughout the study. The LED strips for Sophia and Tatiana were mounted with plastic ties vertically to produce as much light directly to the ports as possible.

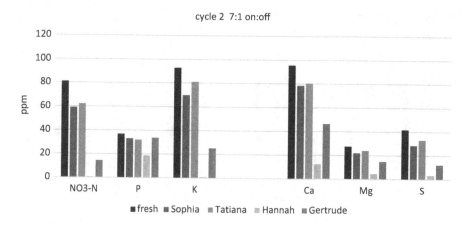

FIGURE 8.26 Nutrient solution ppm concentrations (fresh and at 28 days) for NPK and MgCaS for cycle 2.

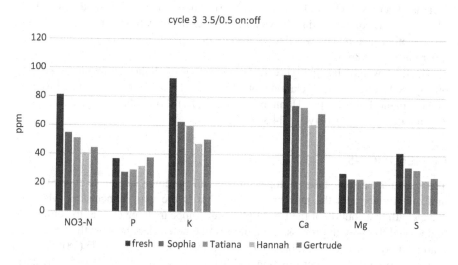

FIGURE 8.27 Nutrient solution ppm concentrations (fresh and at 28 days) for NPK and MgCaS for cycle 3.

The strips for Sophia were 6″ from the ports and for Tatiana they were 12″. It was anticipated that the closer the strips were to the plants that the higher the yields would be. That was not always the case. Also, the light meter was used only at the end of the study, therefore, the intensity may have changed from the beginning to the end of the study.

To better understand the relationship between the PAR value and yields, plots were generated for the second and fourth cycles (7:1 and 4:4) to see whether there was a direct correlation. The plots shown in Figures 8.17–8.24 show that there is

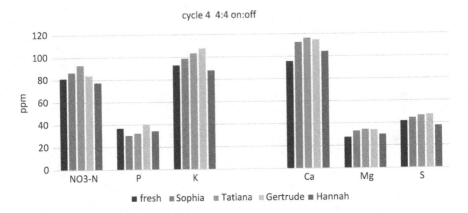

FIGURE 8.28 Nutrient solution ppm concentrations (fresh and at 28 days) for NPK and MgCaS for cycle 4.

some minor correlation for Hannah and Gertrude, a fairly weak correlation for Tatiana, whereas Sophia exhibited an R^2 of 0.62 and 0.69 which is moderate at best. Therefore, it is concluded that the yields are not exclusively based on light energy but involve other factors, i.e., wavelength distribution, plant chemistry, nutrient pH, and other unknown variables.

The fresh weight data in Figure 8.12 is an average of all the mature plants weighed on day 28. Whereas the correlation plots (Figures 8.17–8.24) show individual weights of each plant. For example, Hannah in cycle 2 (7:1) shows plant weights as high as 30 g but the average is between 14 and 15 g.

CONCLUSIONS

These data indicate that the photoperiod plays an extremely large role in plant yields and directly impacts the nutrient uptake. That is the main conclusion. Also, the wavelength distribution of the LED light source appears to be critical to completely understand the growth process. It should also be noted that the growth of kale as it matures may be closer to exponential at the mid-point of its life cycle (28 days), and therefore, time between germination and harvest may skew the results for the early germinators. There were no data collected on the actual time between planting and germination. Most plants germinated in the first seven days of planting. In some cases, plants sprouted after a week or two. The late bloomers were not considered in the average weight calculations, but since the range of weights was broad, it was difficult to determine at what weight the lower threshold should be set for excluding these late bloomers.

FUTURE RESEARCH IDEAS

The research directions for the future cover a wide range of investigations. Each plant needs to be studied individually to determine what is the optimum aeroponic conditions to produce high yields of the plant with the highest nutritional value.

Design of experiment matrices needs to be developed to address all the variables that affect the growth of the plant. One matrix would be the optimum nutrient composition that produces the plants with the highest levels of vitamins and other specific nutrients for that plant. Fundamental to these experiments is understanding the root uptake mechanisms that would allow the roots to absorb the proper inorganic ions. This could involve using the optimum pH, temperature, microorganisms, oxygen, and the nutrient blends. The exposure of the roots to the nutrient solution is another variable that needs to be understood. With the misting and/or the trickle down aeroponic systems, there is a sequence of pump on and pump off. Is there an optimum sequence that allows the roots to take up the nutrients needed by the plant and also become dry enough to absorb oxygen from air? Or does the oxygen need to be dissolved in water to be taken up by the roots? Are simulated drought conditions needed to "encourage" the plant to produce more vitamins or other nutrients that would not occur if the roots are always wet. Also, the photosynthetic rates need to be studied to better understand the optimum conditions for CO_2 concentrations, air temperature, humidity, and the photo energy. This could involve artificial lights in greenhouses or for use in an indoor system—warehouse or container. The optimum lighting also would involve the optimum wavelengths of visible and UV light, the intensity of the light, as well as the photoperiod that the plants are exposed to the light source. The day/night timing sequence would also be a part of these studies. Does each plant have an optimum amount of light exposure to produce the best yield but also the most nutritious plant? Do seedlings require a different light spectrum to germinate and initially grow strong and healthy? Do mature plants need different wavelength combinations? Does the light need to sequenced in the same manner as the sun, i.e., 12 hours on followed by 12 hours off? What if the light was on for 23 hours and off for 1 hour? Does that improve yields and nutrient value of the plant? What about a different rhythm for the plant exposure to light? For example, what affect does 2 hours on 2 hours off have on the yields and nutritional value of a plant. There are hundreds of light sequences that could be studied.

From a practical standpoint, the continuous measure of nutrient uptake (measured by loss of nutrients in the nutrient solution) with selective sensors for nitrate, phosphate, potassium, sulfate, magnesium, and calcium needs to be developed to replace the EC measurements that have been used traditionally for decades in hydroponics but clearly do not provide a clear picture of the nutrients needed by the plant. This would allow for a feedback loop to ensure all nutrients are present at the correct concentration.

There is obviously much work that could be done to better understand the aeroponic process. The field of controlled environment agriculture (CEA) is starting to pick up momentum. The Association of Vertical Farming in Germany is attempting to bring together practioners in the field of hydroponics and aeroponics. That will certainly help to improve the technology. The US Department of Agriculture is also beginning to fund some basic research in this area. There are a few universities that have ongoing research programs in CEA. Some of the key ones have been University of Arizona and Cornell. However, even at the academy, there is very little research being carried out with aeroponics. Most of the research is related to hydroponics and other water culture techniques. Several papers have been published and noted in Chapter 4 from other universities both in the US and around the world. As the chart in Chapter 4 shows the increase in publications, it appears that this trend will continue in the future.

REFERENCES

Ayeni, A., Sciarappa W., Both, A. J., Gurley, T., 2018 unpublished data.
Kozai, T. et al., 2016, *Plant Factory*. Amsterdam: Elsevier.
Kozai, T. et al., 2018, *Smart Plant*. Singapore: Springer.

9 Conclusion

Life is the art of drawing sufficient conclusions from insufficient premises.

Samuel Butler

This book has provided an overview of the state-of-the-art of aeroponic technology from multiple vantage points. It has provided the context which has enhanced the reason why aeroponics has emerged as a new agricultural technology. It has defined what it is and given several examples of aeroponic systems that are functioning today in a variety of settings. These include home use, academic research, business development, innovation, and the practical applications of its principles.

In many ways this technology has benefitted from the work done on the hydroponic front to pave the way for this technology to blossom and become an entity of its own. Both the work done at Disneyworld and NASA highlighted in earlier chapters has certainly been the precursor for much of the research and innovative efforts noted in this book. Without those pioneers, the technology would not be where it is today. The work of the Association of Vertical Farming has been instrumental in bringing together technologists working in area of hydroponics, aeroponics, and the associated technical needs. The Plant Factories in Japan, Taiwan, Korea, and China have also been instrumental in promoting these advances and have contributed to the overall research effort to better define what is optimum. Of course, the growing uncertainty of the future of food production to feed the millions and millions of the next-generation inhabitants of the earth has stimulated interest in exploring this type of technology.

It is noteworthy to see the breadth and depth of the research papers summarized in Chapter 4 on Science. This shows that nascent aeroponics has many more directions that can be pursued but also that it is a technology that has international interest and potential. The exponential increase in the number of technical papers published clearly shows that aeroponics has captured the imagination of many researchers around the world. The scope of plants that can be grown is also highlighted from leafy greens, to vegetables, to potatoes, to trees, and other fruits and vegetables that grow above ground normally or underground. The intellectual property trend is also growing dramatically as new ideas emerge and are documented.

The business success of aeroponics is still in its early phase and it is difficult to predict what the future of this technology compared to hydroponics will be. There are a small number of companies that have started well and seem to be progressing in a positive direction. However, there are also examples of other companies that have appeared and shown promise but eventually have faded away and some have gone out of business. This is normal for any new technological development. The market forces will continue to weed out those technologies that are not competitive and help refine the process to determine which approaches will ultimately survive the test of time. This shaking is a normal effect as new technology emerges and investors evaluate it. One of the biggest questions in the arena of business will be which technology will last and not only produce food at a fair price but also which designs will be the most successful. The two main competing technologies are the horizontal stacked hydroponic approach versus the vertical column configurations. Is it better to do stacking of trays and arrays of LED light or use vertical columns and greenhouses? The economics of these systems will determine which technology survives. Although it is possible that with the need for safe, pure, clean food that human survival will most likely depend on a combination of soil-based agriculture and controlled environment agriculture—hydroponics, aquaponics, and aeroponics—will play a role in various sections of the market. It most likely will come down to the type of plant that is being grown. Some will lend themselves better to one growing method than another. This same process that was experienced with DVD players (beta max vs. VHS) and telecommunication platforms (Blackberry, iPhones, and Android), development will be repeated in the food production arena.

Probably the biggest frontier in the future of aeroponics is the research effort that will need to be exerted to begin to unlock some of the unknowns that were touched on at the end of Chapter 8 on research. It seems that an aeroponic system is uniquely positioned to be able to explore questions about plant physiology and the chemistry of root uptake and light absorption. It can be studied as a completely closed system, where all the inputs can be controlled in a much more comprehensive way than was possible in the area of soil-based agricultural studies. All the variables can be altered and monitored and the desired outcomes can be clearly analyzed and fully understood.

One of the potential trends of this technology is a renewed interest in young people pursuing careers in the agricultural area. Since for years careers in agriculture have been looked down upon as not sophisticated and not economically profitable. Therefore, students have chosen to avoid agricultural science majors. But as the need for a better understanding of the science of aeroponics increases and more and more growers are seen around the country and the world, more students may be inclined to consider a career in agriculture and concomitant fields. One area that will be impacted by this trend is the science of nutrition along with more comprehensive and accurate methodologies to measure nutrition. The current FDA food-labeling practices leave much to be desired. Also, the area of controlled environment agriculture may usher in the capability of producing food with specific concentrations of vitamins, polyphenols, and antioxidants. So instead of a consumer taking vitamin tablets to supplement his deficiencies, he can buy aeroponically grown plants with known levels of these ingredients.

The conclusion of this book is that aeroponics has a very bright future and has the potential to grow much of the safe, clean, pure food that we need to survive as a civilization. The environmental benefits of this technology have been touched on several times throughout this book. The key benefits are the huge savings in water usage; the improvements in productivity, i.e., reduced harvest cycle times; efficient use of nutrients; the possibility of local growing facilities near large population centers; reduced need for pesticides; and the reduction in soil erosion. This technology has already found its way into our homes where growing fresh vegetables and herbs everyday is happening. This trend is most likely to continue and the day may come when much of a family's food supply is being grown in the family greenhouse or garage.

In addition to scaling this technology to the individual consumer, the ideas are proposed by Despommier in his book on Vertical Farming, and we may also see greenhouses appearing on office buildings, near schools and universities, hospitals, retirement villages, and other places where large populations of people are working and living together.

Index

Printed in the United States
by Baker & Taylor Publisher Services

Printed in the United States
by Baker & Taylor Publisher Services